NAME	DATE BORROWED	DATE FOR RETURN
J DARBY	30.11.99	11.1.2000
B. FROST	21.2.01	21.3.01
"	17.4.01	15.5.01
		13.06.01
M J MILLS	28.6.01	26.7.01
	10.9.01	8.10.01
	15.11.01	13.12.01
M. WIPPEBEEK	17.12.01	14.1.02
C TOWNSEND	16.5.02	11.6.02

Multiprotocol over ATM

Multiprotocol over ATM

Building State-of-the-Art ATM Intranets Utilizing RSVP, NHRP, LANE, Flow Switching, and WWW Technology

ANDREW SCHMIDT
DAN MINOLI

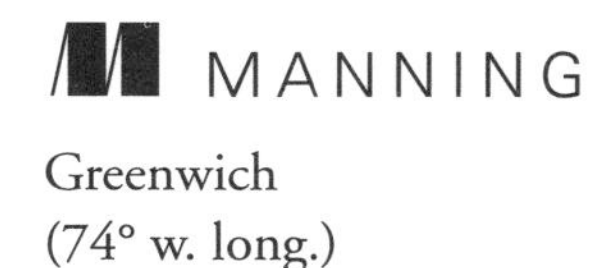

Greenwich
(74° w. long.)

For electronic browsing of this book, see http://www.manning.com

The publisher offers discounts on this book when ordered in quantity. For more information, please contact:

Special Sales Department
Manning Publications Co.
3 Lewis Street
Greenwich, CT 06830

Fax: (203) 661-9018
email: orders@manning.com

♾ Recognizing the importance of preserving what has been written, it is Manning's policy to have the books it publishes printed on acid-free paper, and we exert our best efforts to that end.

Library of Congress Cataloging-in-Publication Data
Schmidt, Andrew.
Multiprotocl over ATM : building state of the art ATM Intranets, utilizing RSVP, NHRP, LANE, flow switching, and WWW technology / Andrew Schmidt and Daniel Minoli.
p. cm.
Includes index.
ISBN 1-884777-42-2 (alk. paper)
1. Intranets (Computer networks) 2. Computer network protocols. 3. Asynchronous transfer mode. I. Minoli, Daniel, 1952– . II. Title.
TK5105.875.I6S36 1997
004.6'6--dc21 97-46424
CIP

Manning Publications Co.
3 Lewis Street
Greenwich, CT 06830

Copyeditor: Katherine Anderson
Typesetter: Dorothy Marsico
Cover designer: Leslie Haimes

Printed in the United States of America
1 2 3 4 5 6 7 8 9 10 – CR – 00 99 98

For Clair and Catherine (AS)
This time, for Gabrielle (DM)

contents

preface

First there was the mainframe, then the desktop, and then the enterprise network. This synthesis has evolved to the point where, as some say, the computer is the network. We are now entering a quantum change phase in the corporate landscape, where the corporation is the network. Since the Industrial Revolution, many, if not most, businesses have been built in the proximity of what can be called *channels of goods distribution.* Initially, cities and businesses were built near harbors or rivers. Later, they were built close to railroads. Still later, they were built within easy reach of the highway system. In the past decade, goods-intensive businesses have been located close to airports.

As we enter the twenty-first century, the corporation is totally defined by its telecommunications network. Today, the network is the channel of goods distribution.

This book focuses on the establishment of corporate intranets using state-of-the-art infrastructures. Technology and business imperatives are advancing rapidly. This book takes a fundamental look at this set of technologies with an eye to educating the corporate planner with regard to the course of action that will best benefit his or her company.

At the computing level, intranets are the culmination of an evolution that started with a movement to client/server architectures in lieu of large mainframe computers. This occurred in the early to mid-1990s. Now companies are starting to actually replace proprietary client/server systems with systems based on World Wide Web (WWW) technology.

A loosely independent but coterminous migration has occurred in the underlying communication infrastructure. By the early 1990s, companies had moved from proprietary communication architectures to Local Area Network (LAN)-based TCP/IP systems. With the emergence of Asynchronous Transfer Mode (ATM) in the mid-1990s, many companies have started to replace the lowest layer of the communication infrastructure, at least at the intranet backbone level, with ATM, so that the communication stack is either IP over ATM LAN Emulation or Classical IP over ATM. During the late

1990s, the protocol stack may further migrate to Multiprotocol over ATM (MPOA) or IP Switching. By the year 2000, the corporate intranet may well look like this: HTTP/TCP/IP/MPOA/ATM

This book is a discussion of how such a network can be built and describes the individual components: WWW, MPOA, LANE, ATM, RSVP, NHRP, MPLS, and so forth. The book is organized in four parts. Part I discusses the computing evolution towards intranet-based technologies. Part II covers start-of-the-art communication approaches, which can be used in support of intranets. Part III covers evolving MPOA-based methods for the deployment of intranets in the later part of the decade. Finally, Part IV addresses alternate technologies to MPOA.

In Part I, chapter 2 provides a description of classical client/server systems. Chapter 3 describes the move towards Internet-based technologies and intranets in particular. Chapter 4 describes the WWW principles and related approaches. Chapter 5 wraps up this discussion with a survey of HTML and related technologies.

In Part II, chapter 6 provides a primer on ATM technologies. Chapter 7 covers the fundamentals of quality of service and how it can be realized on ATM networks. In Part III, chapter 8 sets the ground for the MPOA discussion to follow by describing the integrated services network model and how it applies to quality of service in intranets. Chapter 9 provides a detailed discussion of protocols, such as RTP/RTCP/RTSP, used when building real-time networks. Finally, chapter 10 describes the Resource Reservation Protocol (RSVP).

In Part IV, chapter 11 lays the foundation for the discussion of migrating to an MPOA network by describing the two underlying protocols: LANE and Classical IP over ATM. Chapter 12 then details the MPOA model and its integration with the Next Hop Resolution Protocol (NHRP). Finally, chapter 13 discusses migration strategies used to build MPOA networks and the existing alternative technologies, such as Multi-Protocol Label Switching (MPLS).

acknowledgments

Mr. Schmidt wishes to acknowledge many people at Ameritech, especially Kimberly Price and Tim Waters for their continued support. In addition, special thanks are due to Mark Cnota, Matt Bruening, Brian Stading, and Tim Whiting for their thoughtful consultations while completing this text.

This book is based, in part, on a course given at Stevens Institute of Technology, by Mr. Minoli. Portions of the material (chapters 1-6) are based on class projects and activities. The following individuals are thanked for their input: Peter DePrima, Peter Gaspich, Todd D. Alboum, and David Kinsky. Mr. Minoli also acknowledges the support of R. D. Rosner, Vice President and General Manager, Teleport Communications Group, Internet and Data Services, for the insight, guidance, and assistance provided. Finally, Mr. Minoli thanks Mr. Ben Occhiogrosso, President, DVI Communications, for his suggestions and assistance.

The authors would like to thank the memberships of the ATM Forum and IETF who have contributed tremendous effort in advancing the science of communications.

Finally, the authors would like to acknowledge Dottie Marsico at Manning publication for her contributions to this project.

PART 1

Evolution to Intranets

CHAPTER 1

Introduction to Intranets and Broadband Communications

1.1 Overview and motivation

1.1.1 Where corporations are headed

Since the Industrial Revolution, businesses have built their manufacturing establishments in the proximity of what can be called *channels of goods distribution.* Initially, cities and/or businesses were built in the proximity of harbors or rivers. Later, they were built close to railroads. Still later they were built within easy reach of the highway system. In the past decade, goods-intensive businesses have been located close to airports.

As we approach the turn of the century, the nature of what is considered a manufactured item is changing: either it is simply information (e.g., a cash transfer; a credit given; a reservation established; an analysis, newsletter, or lecture delivered; an electronic order received; a downloaded movie or musical clip; etc.), or it is highly dependent on information (e.g., a just-in-time manufacturing order, a just-in-time inventory transaction, etc.). For a large number of companies, the product is an electronic artifact, which can be called an *e-product*. Electronic commerce in support of e-products, or even real products, is becoming an established way of doing business.

Information Technology (IT) and the supporting communication infrastructure have been evolving since the mid-1960s. First there was the mainframe. Then there was the desktop. Then there was the enterprise network. This synthesis has evolved to the point where, as some say, the computer is the network. We are now entering a quantum change phase in the corporate landscape, where the corporation is the network. Today, the network is the channel of goods distribution. Just as in the past, when access to distribution channels was important, today having a global-reach network is becoming a critical business-survival issue. In this paradigm, the corporation is totally defined by its telecommunication network.

Naturally, that network does not have to be wholly owned by the corporation. The network may be comprised of a *traditional enterprise network* (the physical foundation of the corporate's intracompany communication facilities); the *intranet* (an overlay to the enterprise network which is a way to build uniform applications, clients, and servers having the look and feel of Internet applications); the *Internet*; other *intercompany specialized networks* (e.g., the NYCE banking network); and *international extensions.* It would be desirable if this fundamental synthesis of communication facilities, which we call an *omninet,* would also carry voice, video, image, and other media, in addition to the traditional data objects. For many organizations, omninets start out with support for data or non-real-time multimedia applications, but the long-term goal is to include all forms of corporate communication media.

If the company is the network, companies will have to establish omninets in the immediate future, which will enable them, by the very definition of what an omninet is, to effectively, competitively, easily, and expeditiously trade their e- and pre-e-products, and support the full life cycle of the product development. Also, the omninet allows the company to reach all its affiliated customers at large. These customers at large include buying customers; suppliers; financial companies; insurance companies; local/regional/federal; research laboratories; and employees, including traditional employees, telecommuters, and dispersed temporary consultants.

The place to start on the road to realizing the omninet is with the introduction of intranets in the next couple of years. Electronic commerce and digital cash are expected to be important services in the near future, and intranets are going to be playing a major role, because companies have more control over the transactions when they have an intranet. Employees across the globe have an increasing need to access a company's internal news, corporate policies, training manuals, company directory, and product information. Rather than printing and then shipping materials, an intranet can be used to deliver the information on demand. This saves both time and money. Because of firewall protection, sensitive information can be delivered over an intranet from any number of remote offices.

Many technologies, techniques, and services have evolved in the recent past for designing intranets. In fact, the plethora of such choices is so inclusive that planners may be understandably confused or overwhelmed. The purpose of this book is to shed light on what may be the best alternative to the deployment of such critical, revenue-impacting corporate resources.

Where we are coming from A few years ago, most networks were independent entities, established to serve the needs of a single group, whether in an intraenterprise or interenterprise environment. Users chose hardware technology appropriate to their communication problems. In the past two decades, technology has evolved that makes it possible to interconnect many disparate physical networks and make them function as a coordinated system. The technology, called internetworking technology, accommodates diverse underlying hardware by providing a set of communication mechanisms based on common conventions. The internetworking technology hides the details of network hardware and permits computers to communicate independently of their physical network connections.

In recent years, enterprise networks have been of the internet type. An *internet* consists of a set of connected networks, which act as an integrated whole. The chief advantage of an internet is that it provides universal interconnections, while allowing individual groups to use whatever network hardware is best suited to their needs. The trend, as we enter the first decade of the new century, is towards ubiquitous connectivity. Although internets have been viewed more as a technology (e.g., in terms of routers, bridges, switches, etc.) than actual entities, intranets are being seen as entities that are *syntheses* of technologies to support user *applications*.

There is a large internet *par excellence*. This is the Internet. The Internet with a capital "I" is a massive, worldwide aggregation of subnetworks and computing/information resources located on these subnetworks, including devices such as computers, servers, directories, and so on. It provides a capability for communication to take place between research institutions, government agencies, businesses, educational institutions, and individuals. The Internet includes millions of server computers, housing large volumes of all sorts of information. From 4 million host computers around the world at press time, the Internet is expected to grow to more than 100 million systems in five years. It is estimated that over 20 million users a day in 90 countries access the Internet.

The latest breakthrough in global connectivity began in 1993 with the creation of an Internet subnetwork called the World Wide Web (WWW)—really a software scheme for imposing order over the mass of free form information on the Internet, by organizing it in easily understood pages. What makes the Web such a useful tool is a software technique known as hyperlinking. When composing a Web page, an author can create hyperlinks—words that appear in bold or underscored type and indicate a path to some other information. Using a program known as a Web browser on a PC or workstation, one can read pages stored on any Web computer.

Internet-based technology has now migrated to intranets. Intranets use Internet communication, application, and browsing principles to provide corporate users with uniform access to the company's distributed data repository. These networks are corporate networks, basically portions of or overlays to traditional enterprise networks, which use the same lower layer and application level protocols as the Internet, specifically WWW-related technology. Companies are deploying intranet technology in increasing numbers—for example, Federal Express already has 60 internal Web sites, mostly created for and by employees. The company is equipping its 30,000 office employees around the world with Web browsers, so that they will have access to a plethora of new sites being set up inside the company headquarters. A press-time survey of major U.S. corporations found that 16 percent have an intranet in place, while another 50 percent were considering building one. Market research companies estimated that there were 75,000 servers on the Internet and 75,000 servers on intranets in 1996; they forecast that by the year 2000 there will be 1 million servers on the Internet and 9 million servers on intranets.

Most companies already have the foundation for an intranet. Computers using Web-server software store and manage documents built on the Web's HyperText Markup Language (HTML) format. With a Web browser on the user's PC, the user can call up any Web document, no matter what kind of computer it is on. Firewalls are another element that may already be in place. Intranets allow the presentation of information in the same format to every computer. It is a single system of universal reach. Many client/server systems that will be deployed in the future will be in support of intranets. Broadband communication services will play a significant role in these intranets.

Corporations are seizing the Web as a quick-deployment way to streamline and transform their organizations. These private intranets use the structure and standards of the Internet and the WWW, but they are separated from the public Internet by firewalls. This way, employees can access the Internet, but unauthorized users cannot come into the corporate intranet. Also, access to the Internet can easily be accomplished via the intranet. The Web is an inexpensive, yet effective alternative to other forms of internal communication, in that it provides the mechanism to eliminate paper while increasing accessibility to information. Examples of applications of intranet-based information include internal telephone books, procedure manuals, training materials, and requisition forms. All of this information can be converted to electronic form on the Web and updated in a low-cost manner. To prepare for a meeting an executive could tap into the finance department home page, which has hyperlinks to information such as revenues, forecasts, and so on. Employees can order supplies from an electronic catalog maintained by the purchasing department. Intranets are designed not only to streamline corporate communication, but to offer companies the opportunity to set up proprietary services for their extensive databases.

At the computing level, intranets are the culmination of an evolution that started with a movement to client/server architectures in lieu of large mainframe computers. This occurred in the early to mid-1990s. Now companies are starting to actually replace proprietary client/server systems with systems based on the WWW. Furthermore, observers see an online world turned upside down by Java programming language and the potential for intranets.[1] A brute force approach to get corporate-wide information accessibility would be to put all corporate PCs on the same operating system, which is a daunting, if not impossible, approach. With Java and with an intranet one attains compatibility with any operating system; one can just put a program on an intranet Web server and any employee can download it.

A loosely independent but coterminous migration has occurred in the underlying communication infrastructure. By the early 1990s, companies had moved from proprietary communication architectures to Local Area Network (LAN)–based Transmission Control Protocol/Internet Protocol (TCP/IP) systems. With the emergence of Asynchronous Transfer Mode (ATM) in the mid-1990s, many companies have started to replace the lowest layer of the communication infrastructure with ATM, at least at the intranet backbone level, so that the communication stack is either IP over ATM

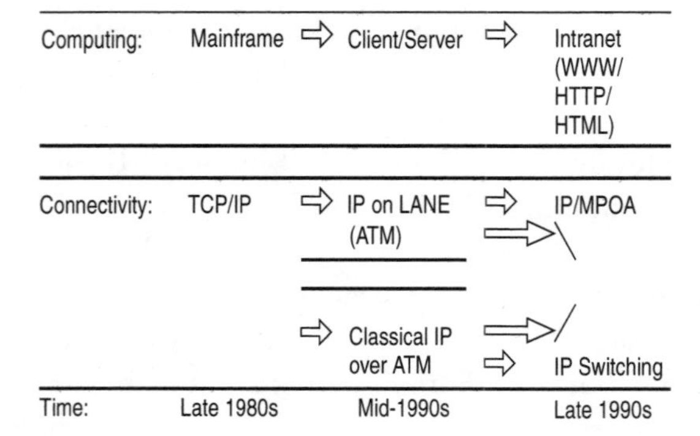

Figure 1.1 Intranets in the late 1990s

LAN Emulation (LANE) or Classical IP over ATM (CIOA). During the late 1990s, the protocol stack may further migrate to Multiprotocol over ATM (MPOA) or IP Switching. See figure 1. By the year 2000, the corporate intranet may well look like this: HTTP/TCP/IP/MPOA/ATM.

What we can look forward to Technology and business imperatives are advancing very rapidly. Since the early 1990s, the Internet-based technology has been sweeping the field of communications. As noted, at this point in time, companies are at the second or third stage of the deployment of the omninet, having established fairly sophisticated enterprise networks, and are exploring both the usage of the Internet and the use of internal intranets. Therefore, this book focuses on the establishment of corporate intranets over state-of-the-art communication infrastructures. Given this rapid evolution, it is not surprising that the literature about the intranet stage is relatively limited. Hence, this book is a discussion of how such a network can be built, as well as describing the individual components: WWW and MPOA.

It discusses one of the more interesting developments in ATM, the MPOA specification, and how network managers can apply the concepts in this specification to build next-generation networks and applications.

Running applications over ATM networks is possible today; however, the techniques of LAN Emulation and Classical IP over ATM are only starting points. In order to take advantage of ATM's potential, new paradigms in network design and in new-product development must be undertaken. MPOA is a product of this paradigm shift and will revolutionize next-generation applications and networks. MPOA can be viewed as solving the problems of establishing connections between pairs of hosts that cross administrative domains, and enabling applications to make use of a network's ability to provide guaranteed quality of service. These changes in paradigms are already beginning to be felt, as manufacturers release products that separate switching from routing, and allow applications to designate their required quality of service.

The MPOA model is the culmination of the work done by several standards bodies in the early to mid-1990s, and it is set up as the specification that will be the basis for networking in the future. With MPOA, hosts are capable of directly communicating with each other, even in the case where the path is between different Logical IP Subnetworks (LISs). LISs are empowered with the ability to communicate without passing data through routers. The MPOA model operates by relying on route servers to maintain knowledge of the location of devices and a network's ability to establish connections that maintain consistent quality of service. When the location is found via the route server, the ATM network can then be used to place a call directly between hosts, and the connection can have an associated quality-of-service specification.

The MPOA specification is critical, because LAN Emulation and Classical IP over ATM suffer from the traditional mindset of segregating networks into subnetworks, interconnected by routers. While this paradigm was the logical transition from bridged interworking, and has worked well for over a decade, it will soon be replaced by techniques that allow clients to communicate directly across administrative domains without using routers. Direct communication means clients and servers will reduce both processing delays and bottlenecks introduced by interconnected routers. Also, and potentially even more powerfully, clients and servers interconnected by ATM switches can utilize ATM's quality of service. The advantages of MPOA are as follows:

- Establishes direct connections for clients to remote servers across the ATM network without using routers
- Uses lower latency in establishing connections between devices
- Takes advantage of ATM direct interdomain connection
- Separates switching from routing
- Provides a unified approach to layer 3 protocols over ATM
- Reduces amount of broadcast traffic
- Optimizes performance by utilizing flexibility in selection of maximum transfer unit size

The work on MPOA has proceeded aggressively within the ATM Forum and it has already specified the protocol model and decided on the interaction between the various components. The ATM Forum has also decided that the LAN Emulation specification would be the foundation of MPOA. Because LANE is providing an emulated layer 2 bridge, it is clearly one of the best techniques for running applications over ATM in a multiprotocol environment.

This book details corporate implementations of MPOA viewed from several perspectives. First, we discuss the network administrator's understanding of the various servers and the basic components of the MPOA solution (i.e., ATM switches and ATM hosts). These can be the same devices that are currently connected to a LANE or Classical IP over an ATM network. The migration to MPOA will involve adding the MPOA servers to the network, which may be as simple as upgrading the ATM switch control software. In all likelihood, basic MPOA bridging will be achievable via software upgrades to LANE bridges.

An additional area where MPOA planners will play a key role is the deployment of applications that utilize quality of service. In order for applications to fully exploit quality of service on an MPOA system, they must be able to communicate their traffic characteristics to the network. This requires a new Application Program Interface (API), along with new applications designed to use the quality-of-service API.

1.2 Setting the stage for creating revenue-enhancing intranets

The intranet, with its reach into the company repository, as well as into the Internet, is designed to offer on-demand access to information via distributed, secure, and platform-independent software objects. The deployment of client/server technology across an enterprise will increasingly require a robust broadband communication infrastructure, such as ATM, which can support the required quality of service and availability of the organization's productivity expectations. Having placed most, if not all, of the employees on an enterprise network, any communication delays, quirks, or inefficiencies can significantly degrade overall company productivity and, ultimately, business viability (naturally, server-level bottlenecks are equally undesirable).

Intranets, with their particular use of client/server technology, necessitate broadband backbones and broadband desktop access. Work functions and activities, viewed from a productivity-enhancing perspective, are better handled via Visual Computer User Interfaces (VCUIs). VCUIs are extensions of the Windows-like Graphical User Interfaces (GUIs) of the late 1980s and of the Mosaic/Netscape-like Network Graphical User Interfaces (NGUIs) of the early 1990s, in that they make extensive use of images, pictures, graphics, diagrams, and virtual reality constructs. In turn, visual-based information requires high communication capacity, both at the local level and at the wide area level. During the rest of the 1990s, more widespread deployment of NGUIs can be expected; at the end of the decade, VCUIs should become prevalent.

Already, large corporations find themselves with a growing need to extend high-speed communications beyond key sites, in order to support applications such as client/server-based company processes (order processing, inventory management, etc.), distributed cooperative computing, business/scientific imaging, video conferencing, video distribution, multimedia, (corporate) distance learning, and so forth. The Internet (and by consequence the intranet) is the platform upon which new communication-enriched applications are now being built. There are two models of computing now evolving to support intranets: a network-centric view, where the computer is the network, or, better yet, the corporation is the network; and a desktop-centric model. We are proponents of the network-centric view. Intranet planners must answer questions such as these [2]:

- What LAN and WAN platforms are best for implementing network-centric applications on both the Internet and intranets? Solutions include 100 Mbps Ethernet (Fast Ethernet), switched Ethernet, switched Fast Ethernet, desktop ATM, and wide area ATM.
- What internetworking strategy is best for implementing network-centric applications on both the Internet and intranets? Solutions include LANE, CIOA, MPOA, and IP Switching.
- What are the best languages and tools for developing reusable objects that make up network-centric applications? Solutions include HTML, VRML, Java, and other extensions.
- What is the best way to extend the capabilities of Web clients?
- What server platform is best for implementing network-centric applications on both the Internet and intranets?

These issues will be covered throughout this book.

Software companies are pushing to market products for the intranet, counting on lower-priced servers and Internet-savvy employees to be a catalyst to the market. Many see intranets as the way for vendors, who, having had little profit on the WWW, may realize a profit from their online services. There is an expectation that consumer-oriented Web business may scale back in the near future, but companies that develop corporate services will become more profitable. Companies dealing in information can have their databases protected behind an intranet firewall and release free server software on the Internet. This software could be downloaded and used to access the database on a paying basis. Important players include Microsoft and Netscape Communications. Compaq, IBM, and Hewlett-Packard were the three largest producers of PC-based Web servers at press time. IBM has positioned Lotus Notes software to tap the market. Microsoft started to sell its Exchange Server in early 1996 and gave away its Internet Information Server software for free over the Internet. Sun Microsystems has released a number of intranet products [1].

The business success of Lotus Notes has pointed to a corporate demand for secure online services. However, the difference between Lotus Notes and an intranet is that Lotus Notes uses a proprietary database structure which replicates data and does not provide quick access to a remote database. An intranet supports internal and external applications, while Lotus Notes is used internally with no gateway to the Internet.

As a further reinforcement of the intranet-like requirements in the corporate environment, online service providers are moving towards making their server software support secure online transactions and services. America Online announced at press time that it would offer a private version of its online services to business customers. In this scheme, America Online plans to allow business customers to buy memberships for their employees, who could then enter a private on-line area of America Online and would be able to access secure email, communicate with coemployees in real time, and access schedule information.

1.3 References

1 G. Gillespie, "Corporations Make Plans to Spin off Their Own Webs," May 1996.

2 J. Flynn, "Battle for the Internet Infrastructure," *Datamation*, May 1, 1996.

CHAPTER 2

Classical Client/Server Environments

2.1 Background

The previous chapter discussed in a broad manner the business imperatives driving corporations to the use of client/server systems and/or intranet/Web-based systems. This chapter discusses the kind of first-generation client/server systems now deployed in many organizations [1–3]. Many principles and approaches have to be taken into consideration when designing second-generation client/server systems, which use intranet/Web methods at the application level and ATM as the network transport technology. This material is summarized from reference [4], which focuses on the support of traditional client/server systems on broadband services such as ATM.

As we venture into this topic, we note that some may take the view that one does not really need ATM as much with client/server outside the server farm, because LAN technologies are very effective with client/server. We view this as being a myopic perspective. The amount of information being transacted across networks continues to increase, particularly as end users become reliant on GUI-, graphics-, and virtual reality-based interfaces. Microprocessors operating at 50 Million Instructions Per Second (MIPS) are common today, and soon processors with capabilities of hundreds of MIPS will be a commodity item (applications requiring 1,000 MIPS are already here). According to Amdahl's Law, a megabyte of I/O capacity is needed for every MIPS of processor performance. Some computing applications already require 1,000 MIPS. This implies that the I/O needs to be in the 50–1,000 Mbps range at this time and higher by the end of the decade. Although the network throughput tends to lag behind the I/O throughput by a few years, these speeds will be needed in the relatively near future. Ethernet only provides a fraction of an Mbps per user (for a typically loaded network), or at most 7 Mbps for a switched system; 100 Mbps Ethernet only provides a few Mbps per user (for a typically loaded network), or at most 70 Mbps for a switched system. The same is true for a Fiber Distributed Data Interface. ATM, on the other hand, provides 155 Mbps *per user*. This is not to imply that FDDI in the local area and services such as frame relay in the wide area are not viable at this juncture. But if the planner is thinking strategically, then ATM is likely the answer. In fact, many corporations have already deployed ATM in the server farm at this time. Furthermore, ATM is a platform that supports both wide area service such as frame relay, as well as legacy LAN services such as Ethernet—hence, ATM will play a significant role in client/server systems one way or another.

2.2 Migration to client/server systems

As discussed in the previous chapter, starting in the late 1980s and continuing through the 1990s, many organizations have transformed their mission-critical computing infrastructures by implementing LAN-based client/server architectures. PCs and workstations are now almost universally connected to LANs. In turn, local or departmental LANs are connected across the corporate enterprise by multiprotocol routers; such tight connectivity is sought whether the organization is clustered in a building or spread out over a campus, metropolitan area, state, nation or several nations, and/or continents. The focus on business accountability in recent years has caused departments and workgroups to introduce workstation technology to solve problems that the traditional central IT organization dealt with slowly or was unable to solve [5].

Companies continue to use automation in general, and IT in particular, to improve productivity of their core business functions. Automation and IT have been employed by companies to "reengineer" the way corporate work is done. For Fortune 5000 companies the recent annual return for their investment in IT has been high [6]. This type of return has fueled the continued introduction of this technology; however, such introduction carries its own cost, which companies are now trying to control. Therefore, IT is itself now being reengineered. Many companies have introduced client/server architectures, and they sometimes take the opportunity of not only reengineering the IT environment, but also accomplishing (some degree of) outsourcing at the same time [3,7,8].

In the 1960s and 1970s, companies looked to their central IT organizations to satisfy evolving market and economic demands. Host-based systems provided adequate support by automating back-room tasks, such as customer information, accounting, and other batch processing jobs. Host-based systems also facilitated control functions and data integrity. The 1980s proved that host-based systems could not keep pace with the growing computing demand imposed on the organization by market and economic forces. Market and economic demands forced departments and divisions to turn to personal computers, workstations, and LANs to achieve the responsiveness they needed to run online mission-critical business applications. End users worked together to achieve competitive advantage and solve their departmental mandates [7].

The PC, workstation, and LAN deployments of the late 1980s and 1990s helped give birth to client/server computing, which efficiently integrates data, applications, and computer services. End users now respond to business demands faster using their desktop and portable computers linked to resources located throughout the network. At the same time, many IT organizations are discovering that they can lower their computing costs by offloading processing from the mainframe to more affordable, smaller machines [7]. While the mainframe provides a level of control over data, the cost savings that organizations can experience are often perceived to be large enough to justify the migration of applications off the mainframe. As a result, IT organizations have moved towards a client/server computing architecture for both cost benefits and flexibility. Figure 2.1 depicts two examples of client/server architectures. The first example shows departmental use of client/server; the second example depicts an enterprise-wide system, which also includes mainframes [9,10].

2.2.1 *Client/server building blocks*

The basics of client/server computing involve dividing the work between two computer systems. One is the system on the desktop, which is generally referred to as the client. The

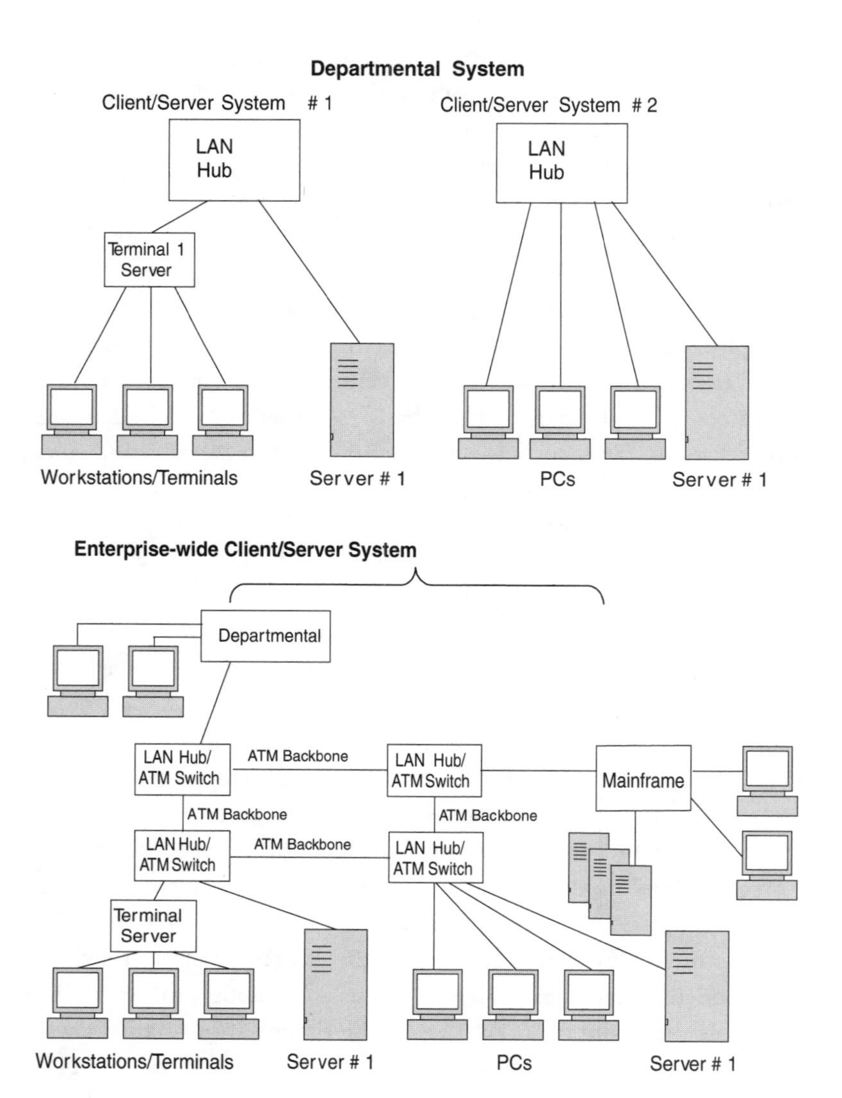

Figure 2.1 Client/server systems: top—departmental system; bottom—enterprise-wide system

other system, the server, is somewhere on the network, along with perhaps other servers. Clients initiate requests for some service—for example, database lookup/update, read from/write to a remote mass storage, dialing out on one of a pool of modems, or printing a report or document; the server accomplishes the task on behalf of the client [5].

The concept behind the client/server architecture is to support a clear separation of functions: the server acts as a service provider, responding to work requests from the various clients. Note that the same piece of equipment can be used as either a client or a server; what makes the difference is the software that is loaded onto the equipment. In fact, the same piece of equipment can act as a client and as a server in different instances. Table 2.1, which is partially based on references [8,11,12], provides, as a working tool, a glossary of key client/server terms and concepts.

The term "client/server" is only loosely defined at the practical level: the term is used to refer to a variety of systems, from simple remote-access file systems to complex database engines. Many people hold that the terms "client/server" and "network database" are synonymous; in fact, the database server technology used in systems developed by Sybase, ORACLE, and others fits most definitions of client/server computing. Some view client/server as the culmination of the trend towards downsizing applications from the minicomputer and the mainframe, to the desktop [8].

The following general definition has been advanced: client/server computing is any application in which the requester of actions or information is on one system and the supplier of actions or information is on another [13]. Most client/server systems have a many-to-one design: more than one client typically makes requests of the server. Many enterprise systems fit this definition.

Table 2.1 Glossary of Key CLient/Server Tems and Concepts

API	The Application Programming Interface (API) is a set of call programs and functions that enables clients and servers to communicate.
Back-End	This is the database engine. Back-end functions include storing, dispensing, and manipulating the information.
Client	This is the (networked) information requester which can query databases and/or other information from a server. At the physical device level, the client is a PC or workstation.
Common Object Request Broker (ORB) Architecture (CORBA)	This is an ORB standard endorsed by the Object Management Group (OMB).
Distributed Computing Environment (DCE)	This is a set of integrated software modules that provides an environment for creating, using, and maintaining client/server applications on a network. It includes capabilities such as: security, directory, synchronization, file sharing, RPCs, and multithreading.

Table 2.1 Glossary of Key CLient/Server Tems and Concepts (continued)

Distributed Database	This is a client/server database system based on a network supporting many clients and servers.
Distributed Relational Database Architecture (DRDA)	An SAA-based enhancement that allows information to be distributed among SQL/DS and DB2 databases.
Dynamic Data Exchange (DDE)	This is a message-based protocol used in the context of Microsoft Windows which allows applications to automatically request and exchange information, thereby supporting interprocess communication. The protocol allows an application running in one window to query an application running in another window.
Front-End	This is a client application working in cooperation with a back-end engine which manipulates, presents and displays data.
Object	This is a named entity combining a data structure with an associated operation. Expressed more formally: Object = {Unique identifier} U {data} U {operations}.
Object Linking and Embedding (OLE)	This is a client/server capability (protocol) in the context of Microsoft Windows that enables the creation of compound documents. It is an extension of DDE's capabilities. A document (e.g., spreadsheet, database entry, video clip, etc.) can be embedded within or linked to another document. When an embedded element is referenced, the application that created it must be loaded and activated. If, for example, the object is edited, these edits pertain only to the document that contains the object and not the more general instance of the object. If the object is linked, it points to an original file external to the document. If, for example, the object is edited, these edits are automatically loaded onto the original document.
Object Request Broker (ORB)	This is software that handles the communication tasks for messages used between objects in a distributed, multiplatform environment.
Object-Oriented Language	This is linguistic support for objects. Objects can be dynamically generated and passed as operation parameters. "Pure" languages offer no other facilities but those related to classes and objects. "Hybrid" languages superimpose object-oriented concepts on an alternative programming paradigm.
Open Software Foundation (OSF)	The OSF is a not-for-profit organization aimed at delivering an open computing environment based on industry standards. It works with members to set technical directions. It licenses software. See also DCE.
OSF/1	OSF's multiprocessing operating system.
Relational Database	This is a database where information access is limited to the selection of rows that satisfy all search criteria.
Server	A server is a computer that stores information that is required by networked clients. It can be a PC, workstation, minicomputer or mainframe.
Windows Open Services Architecture (WOSA)	WOSA is a single system-level interface for connecting front-end applications with back-end services for Windows-based applications. Using WOSA, application developers and users do not need to be concerned about conversing with numerous services, which may employ parochial protocols and interfaces; these details are handled by the operating systems and not by the applications themselves. WOSA provides an extensible framework where applications "seamlessly" access information and network resources in a distributed environment. It uses Dynamic Link Library (DLL) methods to enable software components to be linked at run time.

Even a classical Systems Network Architecture (SNA) fits this definition. The program in the 3270 terminal takes data entered by a user and submits these data (through a cluster controller) to a server application on a mainframe; the server processes the requests and the resulting data are sent back to the terminal. However, when most people talk about client/server they have something else in mind.* Closer to the contemporary use of the word are file server systems such as Novell's NetWare or Banyan's VINES, which respond to client requests by supplying data in various logical aggregations, typically files. Even more complex examples of client/server systems are database engines that support sophisticated data structures. These database systems respond to queries from clients, perform activities based on those queries, and return discrete responses to the clients. In some sophisticated cases, a client/server application can take a complex calculation submitted by a client and partition the calculation into a number of smaller components. These stand-alone calculations are then submitted to a series of calculation servers in a *CPU farm*, which can perform these calculations simultaneously. Once each server has responded with its subresult, the client/server application consolidates these calculation components and is able to rapidly calculate the desired final answer.

2.2.2 Drivers for the establishment of client/server

Based on the previous discussion, a client/server LAN architecture is a computing environment in which software applications are distributed among entities on the LAN. The clients request information from one or more LAN servers which store software applications, data, and network operating systems. The network operating system allows the clients to share the data and applications that are stored in the server, and to share peripherals on the LAN. Figure 2.2 depicts a logical view of a client/server environment.

The client is the entity requesting that work be done; a server is the entity performing a set of tasks on behalf of the client. The client system provides presentation services to the ultimate (human) user. As discussed in chapter 1, there has been an almost universal movement toward GUIs (and more recently VCUIs) as the effective method for presenting information to people. This windowing environment allows the client system (and the user) to support several simultaneous sessions. Facilities such as Dynamic Data Exchange and Object Linking and Embedding (see table 2.1), provide the means to support cut-and-paste operations between such diverse applications as graphics, spreadsheets, and word processing documents.

There are three basic functions supporting computing.

* Keeping a database on a mainframe can be considered consistent with client/server computing principles if the application that uses the database runs on a PC or workstation.

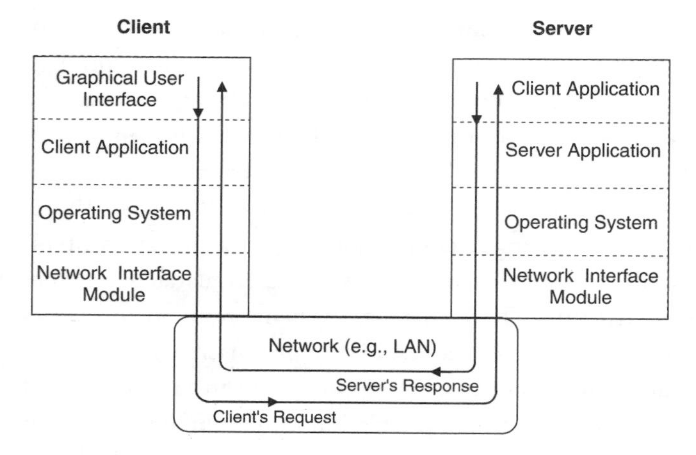

Figure 2.2 Client/server model—logical view

1. Data Management
2. Processing
3. Presentation (to user).

Client/server systems allow the distribution of these functions among appropriate devices, as shown in figure 2.3. Table 2.2 depicts some of the features associated with client/server implementation variations [14, 15].

Table 2.2 Features Associated with Client/Server Implementation Variations

Variation	Application	Examples	Advantages	Disadvantages
Distributed Computing	Computing components executed by most appropriate platform (array processors, mainframe, mini, etc.)	Cooperative Workstations; high-end PCs; minicomputers; mainframes; adjunct processors.	All resources are optimally used	Cooperative computing is required (processing must in some sense be coordinated)
Distributed Database Devices	Resource sharing for desktop applications to scale up	Workstations; high-end PCs; minicomputers; mainframes; Oracle's Server; IBM's Database Manager; Sybase's SQL Server; Novell's NetWare SQL; Gupta's SQLBase; other 4th generation languages	Good utilization of all devices; maximum independence	Difficult to scale up

Table 2.2 Features Associated with Client/Server Implementation Variations (continued)

Variation	Application	Examples	Advantages	Disadvantages
Distributed Presentation	Existing mainframe-based applications	PCs; X-terminals; X-servers;	No changes to existing applications;improved human-machine interface; inexpensive desktop setup	Increased system computational load
Remote Data Access	Common data with independedent applications; decision support systems	Workstations; high-end PCs; minicomputers; mainframes; SQL access; 4th Generation Languages; DECquery	Reliable data; computing choices closer to user	Database machine performance affects all users
Remote Presentation	Independent management of user environment; new or existing applications	Workstations; PCs; DECwindow/Motif	Server is off-loaded;Can be used for existing applications	Degraded performance for multi-application environments

The user's processor controls the user interface and issues commands to direct the activity of the server across the LAN. This is done through the use of Remote Procedure Calls (RPCs). RPCs are software programs with distributed capabilities. Applications

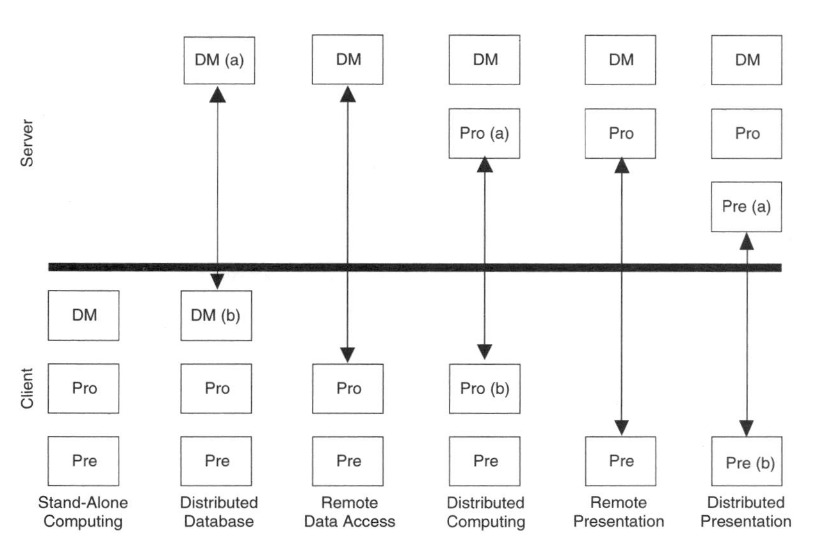

Figure 2.3 Various client/server implementations

that are implemented on the LAN can call these procedures by ordering messages, translating different codes, and maintaining the integrity of the protocol. Not all applications in a client/server architecture are stored on a server: clients may also be capable of storing applications and data locally. When clients possess individual operating systems, the network is referred to as loosely coupled.

The goal is to have a client/server platform that operates in an open environment. In an open environment, the requester-service procedures and necessary support messages are based on well-defined standards, allowing multiple platforms (possibly using different hardware and software) to interact seamlessly. RPCs play a role in the need for openness. One of the key advantages of an open environment is that servers may grow as the need arises, without being forced to buy from a single supplier. Additionally, any needed changes to the operating system or hardware can be done without having to modify the client applications. Some of the benefits of using a client/server architecture include the following [14]:

- Increased productivity
- Control/reduction of costs by sharing resources
- Ease of management through focusing efforts onto a few servers
- Ability to adapt to new needs.

In the stand-alone computing environment, all the intelligence is placed in the PC. Even when the data reside in a file server, the file server simply stores the file in a functionally equivalent manner as if it were a local hard disk—for example, to do a sort, the entire file is downloaded to the PC, which does all the computing. The file may eventually be reloaded in a sorted form to the server. The server has no ability to manage or control the data [16].

In a more sophisticated client/server environment, the server has the capability to perform database management. This means that the server (also known as the *back-end* or *database engine*) can run a (relational) database management system using a multitasking operating system (e.g., OS/2 or UNIX).

A relatively powerful microprocessor is required to run the multitasking operating system, the database manager, and the concurrent user sessions. Typically, this is a Pentium-based system.

In a client/server environment, the workstation (also known as the *fron-end*) is responsible for presentation functions (i.e., display of data according to specified user interfaces, editing and validating data, and managing the keyboard and mouse). These functions are easy to implement, making the cost of the repetitive module (N modules for N end users) low. This results in the following advantages [16]:

- Support of many concurrent users (a few hundred)
- Reduction in cost, since front-end software is simple
- Availability of software from a variety of vendors, performing a variety of functions and sharing the database
- Data integrity (single database shared by all users).

2.2.3 Client/server linking

These are the four common methods of linking client/server applications:

1. APIs
2. Database servers
3. Remote windowing
4. RPC software

APIs A commonly used method of sharing information on a client/server network is through the use of network APIs. These are vendor-provided functions that enable application programmers to access network resources in a standardized manner. Network APIs also allow for the connection of various applications running under the same operating system. Vendor-supplied APIs are not easily portable among different operating systems. SPX (Novell NetWare), Named Pipes (Microsoft LAN Manager), and Sockets (Berkeley UNIX) are examples of APIs. A long-term goal is to come up with vendor independent APIs.

Database servers A database server is a dedicated server in a client/server network that provides clients with distributed access to database resources. The database servers usually employ a Structured Query Language (SQL) relational database to communicate between the client and the server. SQL is a de-facto standard language supported by many vendors, which enables it to run on multiple platforms (a number of vendors have some syntactical differences and extensions meant to improve developer productivity—these extensions should be assessed to determine if the benefit accrued from their use outweighs the incompatibilities that may occur). Using SQL, database servers allow a client to download a table, as opposed to downloading the entire database. This feature reduces LAN and WAN traffic, thereby improving performance. SQL database servers extend user processing applications across various network operating systems via RPCs. SQL is a simple, conversation-like language relying on English commands such as SELECT, FROM, WHERE, and so on, to perform database inquiries.

Remote windowing Remote windowing is an extension of the windowing concepts commonly used with PCs. Remote windowing allows for multivendor connectivity through the use of the Universal Terminal Standard. This concept allows the viewing of multiple user applications concurrently from remote locations. In order to properly distinguish Microsoft Windows from the generic concept of *windowing*, the following definition is provided. Windowing is a software feature that offers split-screen capability in which the different partitions of the display form rectangular areas. These rectangular forms, usually accompanied by GUIs, can be moved or resized on the screen.

RPCs RPCs are based on Computer-Aided Software Engineering (CASE) principles. This concept allows conventional procedure calls to be extended across a single- or multivendor network. RPCs function across several communication layers. Programmers are shielded from the networking environment, allowing them to concentrate on the functional aspects of the applications under development. The computing industry generally describes two RPC technologies as industry standards. They are: Open Network Computing (ONC) RPC (often called SunRPC), and Network Computing System (NCS). An international standard RPC would be desirable.

2.2.4 Client/server services

Client workstations operate by issuing requests for services (see table 2.3 for a typical list). The responder to such service may be on the same processor (e.g., request for a locally stored file), or a remotely networked processor. The critical fact here is that the format of the request must be the same, regardless of where the responder is located. The NOS intercepts the request, and, if necessary, translates or adds the details required by Typical Client/Server Services the intended responder. This NOS service is known as redirection. This capability intercepts the client's operating system calls and redirects them to the server operating systems. Requests such as file access, directory inquiries, printing, and application activation, are trapped by the redirection software and redirected over the LAN or WAN to the correct server.

Table 2.3 Typical Client/ServerServices

Database Services	First-generation database services were actually file server functions with an altered interface, and execute the database engine mostly in the client. More sophisticated systems support database interactions via SQL syntax (e.g., Sybase, IBM's Database Manager, ORACLE, Ingress, etc.).
Fax Services	Clients can route requests for faxing, even though the fax system may be busy with another document. Requests are redirected by NOS software and managed in a fax queue. Notification of delivery is often provided. Applications themselves need not provide these capabilities.

Table 2.3 Typical Client/ServerServices (continued)

File Services	File Services support access to the virtual directories and files located on both the client's hard disk and in some network server (here through redirection).
Message Services	With this service, which supports buffering, scheduling, and arbitration, messages can be sent to and received from the network.
Network Services	These services provide support for actual communication protocols such as TCP/IP (over Ethernet, Token Ring, FDDI, etc.), APPN, and so on. These services enable LAN and WAN connectivity. The application accesses the network services over the API.
Print Services	Clients can route requests for printing, even though the printer may be busy with another document. Requests are redirected by NOS software and managed in a printer queue. Notification of delivery is often provided. Applications themselves need not provide these capabilities.
Remote Boot Services	The skeleton operating system is contained in the local workstation (e.g., on an x-terminal); its capabilities are loaded over the network.
Windowing Services	This service provides the workstation with the ability to activate, rescale, move, or hide a window. Applications themselves need not support these physical windowing capabilities (the application, using the GUI capability, places data into a virtual screen—the window service handles screen placement and manipulation).

As seen in table 2.3, application servers provide the same business functionality as was supported in the past by (more expensive) mainframes and minicomputers. These services are invoked in response to a client's request, specifically via RPCs. As already indicated, these requests are issued by the client system to the NOS. The request processing module of the client formats the request into the appropriate RPC and passes that to the application layer of the client's application layer of the communication protocol stack. What now becomes a Protocol Data Unit (PDU) is further passed down the protocol stack, sent over the network, and received by the server through the communication protocol stack. The PDU is reconstituted into the RPC, which is delivered to the server's processing logic.

RPCs standardize the way programmers write calls (invocations) to procedures (e.g., subroutines) stored somewhere in the network. If an application issues a functional request and this request is embedded in an RPC, the requested function can be located anywhere in the enterprise network (or even outside the company's enterprise); naturally, considerations regarding security and access rights have to be taken into account. The RPC facility provides the invocation and execution of requests from processors running on possibly different operating systems and different hardware platforms. Many RPCs also provide data translation services; here the call causes dynamic translation of data between systems with different physical data storage formats. RPC standards are evolving and are being adopted by the industry in order to promote open environments [8,13].

2.2.5 Distributed Computing Environment

A Distributed Computing Environment (DCE) is a set of integrated software modules that provides an environment for creating, using, and maintaining client/server applications on a network (see [1] for a more extensive discussion). DCE is sponsored by the Open Software Foundation (OSF), which is a not-for-profit organization aiming at delivering an open computing environment based on industry standards. Some see DCE as the most important architecture to be defined for the client/server technology. DCE includes capabilities such as security, directory, synchronization, file sharing, RPCs, and multithreading. One can view DCE as a "black (software) box"* installed by the (major) hardware and software vendors that in theory eliminates technological barriers. Such black (software) boxes connect a variety of operating systems, dissimilar hardware platforms, incompatible communication protocols, and application and database systems in a transparent manner for all concerned (end users, system managers, and application developers) [8]. Another way of looking at this is as a bridge between the embedded base of applications and future applications and platforms.

2.2.6 Benefits of client/server

Organizations, driven by the pressures for IT cost-reductions, properly inquire if a client/server-based solution is the correct approach to their application development needs. There are advantages in deploying client/server applications. Client/server systems typically deliver more information to the user; they do this faster and potentially in a more readable format than is available with legacy systems [10,13]. Many client/server applications can be built using off-the-shelf tools that take advantage of GUIs.

Client/server systems enable the users to be more productive through the integration of personal desktop tools and corporate systems. Mechanisms such as OLE, DDE, and CORBA (see table 2.1) support these productivity improvements. Client/server systems leverage the business experience of the user in combination with tools that are usable by noncomputer professionals.

Client/server applications usually result in (long-term) monetary savings, although these savings can only be achieved if the cost to support distributed applications is low. Studies have shown that when mainframe-based legacy systems are replaced with client/server systems, their combined costs are, in most cases, less than the operating system, application software, hardware, and maintenance costs of the original system. Cost

---

* More precisely, it is a prepackaged set of integrated interoperability applications (RPCs, presentation services, naming, security, threads, time services, distributed file services, management, and communications) which enables connection of diverse hardware and software systems, applications, and databases.

reductions are also realized by more efficient use of the computing resources. In a client/server system, the client only formulates the request for data and then processes the reduced data set returned by the server. This implies that the user requires a less powerful processor than if the client performed all the application tasks locally. In fairness, however, the cost of machine cycles has been coming down, so that these savings alone may not be as important as time goes by.

The implementation of a GUI-based presentation front end to an application (as shown in figure 2.3) can also result in savings in terms of a shorter learning time for new users and, perhaps, a shorter time to complete a given production task, particularly when shortcuts are implemented, obviating the need to navigate multilayer menus. The IT organization can deliver data to end users faster in the form they want because end users choose their front-end tools and IT organizations maintain central control over the data with programmable servers and other tools [7].

Additionally, a well-designed client/server query/response process results in less data being transacted over the network. This eliminates (or reduces) the cost of having to incrementally upgrade the network components to provide additional bandwidth, particularly in a WAN context where bandwidth costs are still nontrivial (this, however, also applies at the local level, pushing the user to higher speed LANs—e.g., 100 Mbps LANs, FDDI, and ATM LANs). This upgrade also applies to bridges, routers, modem pools, and so forth. However, the requirement for bandwidth is inevitably on the increase, and ATM will be needed at some point in the next few years in many companies. Network management tools are essential to control traffic in the client/server system. Also, coming changes in networking options make it critical that applications are built using standard protocols to isolate the location of the data and process the application. Thus, a database or application server may move from the department to the headquarters as new high-capacity communication services (such as ATM) reduce the communication charges (due to economies of scale and integration) and provide LAN-like performance over the WAN.

Not all client/server systems now being installed, however, aim at replacing legacy systems. IT managers realize that in many cases these applications can be used to enhance the capabilities obtainable with existing systems.

Many organizations have initially seen client/server emerge in a grass-roots fashion, at the departmental level. In these organizations, success in departmental client/server computing is now driving an enterprise-wide demand for greater access to corporate data, for access to applications once in the hands of a few users, and to email services, in order to further satisfy departmental requirements for improving business productivity and remaining competitive. However, client/server computing on a departmental level only provides benefits within specific departments; as a result, data are duplicated on other servers or PCs by rekeying or downloading information from the mainframe (new data may also be created completely outside the mainframe environment on a network server). The desirable solution is to balance end-user responsiveness with IT control by deploying client/server computing across the entire organization. Proponents hold that client/server computing is now proving itself as an effective architecture of strategic value for linking all of an organization's computing resources, in the form of enterprise-wide client/server computing as shown in figure 2.1 [7]. In this computing environment, endusers, departments, and IT can merge to create one cohesive computing environment. Front- and back-office end users anywhere in the organization can now access corporate data on their desktop, and central control of data is put back in the hands of the IT department.

2.2.7 Rollout phases

Enterprise-wide client/server computing empowers organizations to reengineer business processes and distribute transactions in order to streamline operations, while providing new and better services to customers. The transition to enterprise-wide client/server computing is an evolutionary effort with three steps (these steps have been used by major organizations that have already made the transition) [7]:

1. Deployment of client/server computing for departmental applications.
2. Integrating mainframe-based applications with a client/server network.
3. Deployment of client/server computing on an enterprise-wide level.

(Naturally, some progressive companies can go directly to the third step if they choose; however, the three-phase evolution is more common.)

Deployment of client/server computing for departmental applications As noted earlier, client/server computing first evolved within workgroups and departments in order to cut costs, provide alternatives to host systems, and improve performance and access to information. Client/server computing can provide a substantial competitive advantage by enabling departments to work faster and better. Client/server computing can create, when properly implemented, a more effective environment for line-of-business applications.

Integrating mainframe-based applications with a client/server network The successful deployment of client/server systems on a departmental level sets the stage for the next phase: end-users soon seek access to data outside their existing client/server environment. As the integration occurs, the client/server system interoperates with relational and nonrelational database management systems, indexed files, mainframe data,

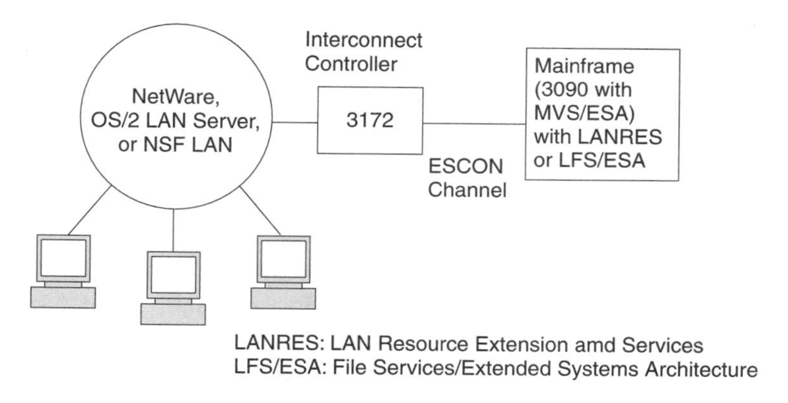

Figure 2.4 Integrating a mainframe as a superserver

and other services provided by existing legacy applications. This can be accomplished through gateway products and open interfaces. Mainframe vendors, such as IBM, are beginning to bring out mainframe-LAN software to facilitate the use of mainframes as superservers in client/server networks. Figure 2.4 shows an example [17]. Users are beginning to create server farms in secure rooms and letting the IT department manage them; a mainframe-based alternative to a multitude of servers may be cost effective in specific situations.

Deployment of client/server computing on an enterprise-wide level With the establishment of successful departmental client/server networks that interoperate with resources outside those networks, organizations are ready to pursue the most strategic stage of client/server development: enterprise-wide client/server computing. This phase goes well beyond departmental client/server and gateway products by providing the total integration of department and corporate IT applications that span the enterprise. This environment enables the organization to leverage both existing central and line-of-business systems. With enterprise-wide client/server applications, the IT organization is able to reestablish control over data, while at the same time supporting a truly distributed environment. The IT organization can now maximize the value of information by increasing its availability and at the same time maintain central control over data integrity. End users can access data from anywhere in the enterprise from their desktops.

2.3 *Communication systems in place for client/server systems*

Client/server systems rely, by definition, on communication and networks. Client/server computing can be implemented on a number of communication protocol suites, including NetWare IPX/SPX, NetBIOS, TCP/IP RPC, OSI, and SNA LU6.2. IT managers must select the protocol suite that best suits their installed base and applications inventory. One advantage of TCP/IP is that many (if not the majority of) applications support the Internet protocol suite. There is an increased need for multiplatform connectivity, making the problem technically challenging. Early SQL versions often used a considerable portion of the network's bandwidth by pushing nonessential data onto the network; later SQL versions from Oracle and Sybase, among others, make better use of bandwidth by storing procedures on the server. This reduces the amount of request data required from a server to trigger a transaction [10]. Developing client/server applications across a WAN is even more challenging.

To make distributed computing a reality, one needs a sophisticated network infrastructure which can support not only client-to-server communication, but also server-to-server communication [18]. Server-to-server interactions are complex and resource-intensive. Some distributed database management systems support data replication, which is the process of copying data, distributing it to remote sites, and retaining synchronization as changes are made to the data in real time. However, without an effective underlying network (i.e., a network of adequate bandwidth, reliability, flexibility, etc.), replication is of limited value. The need to establish flexible WAN infrastructures is now recognized by many of the network database vendors [18].

This section examines communication systems at the local and remote level that may be used to support client/server environments. Except for this section, the discussion in the rest of the book focuses on ATM. Readers familiar with these topics may choose to skip over the following synopsis.

2.3.1 *Local area connectivity*

Local connectivity for traditional client/server systems or intranet based systems is physically achieved using LANs. Strategically, higher-speed LANs will be required; however, in the short term any number of technologies will probably be adequate to various degrees. LAN technology has encompassed three generations. First-generation technology spanned the period covering the mid-1970s to late 1980s. Many corporations have or are still deploying these LANs based on coaxial cable or twisted-pair cable media. Second-generation technology emerged in the late 1980s to early 1990s, and is based on shared fiber-cable media or high data-rate copper; this includes FDDI and 100 Mbps Ethernet. Third generation LANs are now beginning to become commercially available, and will see major deployment in the late 1990s to support new high-bandwidth applications such as multimedia, video, desk-to-desk video conferencing and high-throughput client/server systems.

The transmission speed achievable with LANs varies from 4 to 16 Mbps for first-generation LANs; 100 Mbps for second-generation LANs; and Gbps rates for third-generation LANs. The need to interconnect collocated and/or remotely located LANs has emerged as a key need of the 1990s. The trend is towards enterprise-wide networking, where all departments of a company are interconnected with a seamless (backbone) network, allowing company-wide access to all information and hardware resources. Enterprise-wide client/server systems rely on these enterprise-wide networks.

LAN technologies First generation LANs were developed in the early 1970s to provide what was then considered high-speed local connectivity among user devices. The contention-based Ethernet LAN technology was brought to the market by a joint effort among Xerox, Intel, and Digital Equipment Corporation. Ethernet initially employed coaxial cable arranged in a logical bus and operating at 10 Mbps; now, thin coaxial and twisted-pair cable are used, usually in conjunction with wiring hubs and physical star arrangements. Extensive standardization work has been done by the Institute of Electrical and Electronics Engineers (IEEE) in the past 15 years, leading to well-known standards such as the IEEE 802.2, 802.3, 802.4, and 802.5.

In the early 1980s, a token-bus and a Token Ring technology were also developed, operating at 4 Mbps (a 16 Mbps system is also available). The token medium-sharing discipline is a variant of the polling method common in traditional data networks; however, instead of centrally controlled polling, the token is passed from station to station in an equitable manner. Only the LAN station possessing the token can transmit. Token Ring systems took the approach of using (shielded) twisted-pair wires as the underlying medium, mainly because such a medium is cheaper and simpler to install than coaxial cable (however, Unshielded Twisted-Pair (UTP) is expected to become the dominant LAN medium for traditional LANs). Over the past decade, the cost of connecting a user to a LAN decreased from about $1,000 to less than $50.

Higher network performance is required in order to support the applications now being put online by organizations. These applications go beyond the movement of just data and include imaging, multimedia, and video conferencing. One way of increasing the bandwidth available to applications is to replace the existing network with one based on FDDI.

Efforts on second-generation LANs started in the early 1980s; products began to enter the market in the late 1980s. This token-based backbone campus technology extends the features of LANs in terms of geographic radius, now covering a campus, as well as speed, now reaching 100 Mbps. Implementers initially settled on multimode fiber as the underlying medium, although support for single mode fiber was added in the late 1980s. One factor that has slowed down the deployment of FDDI systems has been the cost of the interface cards. The cost of connecting a user to an FDDI LAN started out at about $8,000 and is now around $500 to $1,200. Efforts to facilitate the use of twisted-pair copper wires for FDDI have been underway, in order to bring the station access cost down (copper-based interfaces cost between $500-700). While standards work in this area has been slow, progress has been made in the recent past. Ethernet technology at a speed of 100 Mbps is now being deployed, as well as higher speed Token Ring technology. Both of these technologies are candidates for second-generation systems.

Starting in 1990, efforts have been underway to develop third-generation LANs supporting gigabyte-per-second speeds (0.2–0.6 Gbps) over UTP or fiber facilities [19]. These efforts are based on ATM mechanisms. ATM mechanisms were first developed in the context of wide area networking; the same technology is being applied in the premises networking context using ATM-based hubs and switches. Work along these lines is sponsored by industry vendors under the auspices of the ATM Forum. ATM switches to support high-end workstations have been available commercially since 1993. Workstation manufacturers are developing interface cards to connect their equipment to ATM switches. Initial costs were around $4,000 per port, but these costs should come down considerably (to $700) in the next couple of years, as chipsets emerge (currently the cost for a desktop connection is as low as $250 for 25 Mbps LANs and $750 for 155 Mbps LANs).

LAN topologies There are three major physical (first-generation) LAN topologies: star, ring, and bus. A star network is joined at a single point, generally with central control (such as a wiring hub). In a ring network the nodes are linked into a continuous circle on a common cable, and signals are passed unidirectionally around the circle from node to node, with signal regeneration at each node. A bus network is a single line of cable to which the nodes are connected directly by taps. It is normally employed with distributed control, but it can also be based on central control. Unlike the ring, however, a bus is passive, which means that the signals are not regenerated and retransmitted at each node.

Other configuration variations are available, particularly when looking at the LAN from a physical perspective: the star-shaped ring and the star-shaped bus. The first variation represents a wiring methodology to facilitate physical management: At the logical level the network is a ring; at the physical level it is a star, centralized at some convenient point. Similarly, the second variation provides a logical bus, which is wired in a star configuration using wiring hubs. Table 2.4 summarizes the use of these topologies in the three generations of LANs.

Table 2.4 LAN Topologies

LAN	Early	Recent
1st Generation, Broadband	Bus	Bus
1st Generation, Ethernet	Bus	Star-shaped bus
1st Generation, Token Ring	Ring	Star-shaped ring
2nd Generation	Fiber Double Ring	Star-shaped double ring
3rd Generation	Star-based access	—

Medium-sharing disciplines As discussed, in traditional LANs there are two common ways of ensuring that nodes gain orderly access to the network, and that no more than one node at a time gains control of the shared LAN channel. The first is by the contention method; the second is by the token variant of polling.

The contention method is known as Carrier Sense Multiple Access with Collision Detection (CSMA/CD). If a node has a message to send, it checks the shared-medium network until it senses that it is traffic free and then it transmits. However, since all the nodes in the network have the right to contend for access, the node keeps monitoring the network to determine if a competing signal has been transmitted simultaneously with its own. If a second node is indeed transmitting, the two signals will collide. Both nodes detect the collision, stop transmitting, and wait for a random time before attempting to regain access.

Token-based LANs avoid the collisions inherent in Ethernet by requiring each node to defer transmission until it receives a token. The token is a control packet which is circulated around the network from node to node, in a preestablished sequence, when no transmission is otherwise taking place. The token signifies exclusive right to transmission, and no node can send data without it. Each node constantly monitors the network to detect any data frame addressed to it. When the token is received by a node, and the node has nothing to send, the node passes it along to the next node in the sequence. If the token is accepted, it is passed on after the node has completed transmitting the data it has in its buffer. The token must be surrendered to the successor node within a specific time, so that no node can monopolize the network resources. Each node knows the address of the predecessor and the successor. Note that ATM-based networks are not subject to this kind of contention.

Lower layer LAN protocols In a LAN environment the functions of layer 1 and 2 of the OSI reference model have been defined by the IEEE 802 standards for first-generation LANs, ANSI X3T9.5 for second-generation LANs, and industry groups such as the ATM Forum, the Exchange Carriers Standards Association, and the International Telecommunications Union (the last two bodies having standardized the supporting ATM functions) for third-generation LANs.

Using the networking protocols defined at layer 3 (such as IP—Internet Protocol) and connection-oriented transport layer protocols (such as TCP—Transmission Control Protocol), one can build the LAN protocol suite up to layer 7 in order to support functions such as email, file transfer, directory, and so forth. TCP/IP has often been used commercially.

Because LANs are based on a shared medium, the data link layer is split into two sublayers. These sublayers are the Medium Access Control (MAC) and the Logical Link Control (LLC). The LLC sublayer provides a media-independent interface to higher layers. The MAC procedure is part of the protocol that governs access to the transmission medium. This is done independently of the physical characteristics of the medium, but takes into account the topological aspects of the subnetwork. Different IEEE 802 MAC standards represent different protocols used for sharing the medium (IEEE 802.3 is the contention-based Ethernet and IEEE 802.5 is the token-based system [see table 2.5]).

Table 2.5 Functions at Specified Protocol Levels

LLC	• Reliable transfer of frames • Connection to higher layers
MAC	• Addressing • Frame construction • Token/collision handling
PHY	(Physical Layer Protocol—explicit only in more recent standards such as FDDI and local ATM) • Encoding/decoding • Clocking
PMD	(Physical Medium Dependent—explicit only in more recent standards such as FDDI, SONET, and ATM) • Cable parameters (optical/electrical) • Connectors

Connectionless versus connection-oriented communication

Two basic forms of communication (service) are possible for both LANs and WANs: connection-oriented mode and connectionless mode.

A connection-oriented service involves a connection establishment phase, an information transfer phase, and a connection termination phase. This implies that a logical connection is set up between end-systems prior to exchanging data. These phases define the sequence of events, ensuring successful data transmission. Sequencing of data, flow control, and transparent error handling are some of the capabilities inherent with this service mode. One disadvantage of this approach is the delay experienced in setting up the connection. Traditional carrier services, including circuit switching, X.25 packet switching, and early frame relay, are examples of connection-oriented transmission; LLC 2 is also a connection-oriented protocol.

In a connectionless service, each data unit is independently routed to the destination. No connection establishment tasks are required, since each data unit is independent of the previous or subsequent one. Hence, a connectionless mode service provides for transfer of data units (cells, frames, or packets) without regard to the establishment or maintenance of connections. The basic MAC/LLC (i.e., LLC 1) transfer mechanism of a LAN is connectionless. In this connectionless mode, transmission delivery is uncertain, due to the possibility of errors. Connectionless communication shifts the responsibility for the integrity to a higher layer, where the integrity check is done only once, instead of being done at every other layer. (Naturally, protocols other than IP can be used.)

TCP/IP protocol suite The basic TCP/IP protocol suite is shown in figure 2.5 for both LAN and WAN environments (there are about 100 protocols in the Internet suite). A TCP/IP LAN application involves a user connection over a standard LAN system (IEEE 802.3, .4, .5 over LLC); software in the PC and/or server implementing the IP,

	LAN Environment	**WAN Environment**
Layer 7–Layer	Application-specific protocols such as TelNET (terminal sessions), FTP and SFTP (file transfer), SMTP (email), SNMP management), and DNS (directory)	
Layer 4	TCP, UDP, EGP/IGP	TCP, UDP
Layer 3	IP, ICMP< ARP, RARP	IP, ICMP X.25 PLP
Layer 2	LLC, CSMA/CD, Token Ring, Token Bus	LAP-B
Layer 1	IEEE 802.3, .4, .5 (PMD Portions)	Physical Channels

Note:

SFTP = Simple File Transfer Protocol	FTP = File Transfer Protocol
SMTP = Simple Mail Transfer Protocol	SNMP = Simple Network Management Protocol
DNS = Domain Name Service	UDP = User Datagram Protocol
ICMP = Internet Control Message Protocol	ARP = Address Resolution Protocol
RARP = Reverse Address Resolution Protocol	EGP = External Gateway Protocol
IGP = Internal Gateway Protocol	PMD = Physical Medium Dependent

Figure 2.5 TCP/IP-based communication: key protocols

TCP, and related protocols; and programs running on the PCs and/or servers to provide the needed application (the application may use other higher-layer protocols for file transfer, network management, etc.).

In a TCP/IP environment, IP (specifically, version 4—IPv4) provides the underlying mechanism to move data from one end system (say, the client) on one LAN to another end system (say, the server) on the same or different LAN. IP makes the underlying network transparent to the upper layers, TCP in particular. It is a connectionless packet delivery protocol, where each IP packet is treated independently (in this context, packets are also called *datagrams.*) IP provides two basic services: addressing and fragmentation/reassembly of long packets. IP adds no guarantees of delivery, reliability, flow control, or error recovery to the underlying network other than what the data link layer mechanism already provides. IP expects the higher layers to handle such functions. IP may lose packets, deliver them out of order, or duplicate them; IP defers these contingencies to the higher layers (TCP, in particular). Another way of saying this is that IP delivers on a *best-effort basis*. There are no connections, physical or virtual, maintained by IP.

Since IP is an unreliable, best-effort connectionless network layer protocol, TCP (a transport layer protocol) must provide reliability, flow control and error recovery. TCP is a connection-oriented, end-to-end reliable protocol providing logical connections between pairs of processes. Some TCP features include the following.

- Data transfer: From the applications viewpoint, TCP transfers a contiguous stream of octets through the interconnected network. The application does not have to segment the data into blocks or packets since TCP does this by grouping the octets in *TCP segments*, which are then passed to IP for transmission to the destination. TCP determines how to segment the data and it forwards the data at its own convenience.
- Duplex communication: TCP provides for concurrent data streams in both directions.
- Flow control: The receiving TCP signals to the sender the number of octets it can receive beyond the last received TCP segment without causing an overflow in its internal buffers. This indication is sent in the acknowledgment in the form of the highest sequence number it can receive without problems (this approach is also known as a window mechanism).
- Logical connections: In order to achieve reliability and flow control, TCP must maintain certain status information for each data stream. The combination of this status, including sockets, sequence numbers and window sizes, is called a *logical connection* (also known as virtual circuit).
- Multiplexing: This is achieved through the use of a ports mechanism.

- Reliability: TCP assigns a sequence number to each TCP segment transmitted and expects a positive acknowledgment from the receiving peer TCP. If the acknowledgment is not received within a specified timeout interval, the segment is restransmitted. The receiving TCP uses the sequence numbers to rearrange the segments if they arrive out of order and to discard duplicate segments.

(Naturally, protocols other than TCP can be used.)

FDDI The FDDI is a set of standards that defines a shared-medium LAN utilizing fiber (single mode and multimode) and, now, twisted-pair cabling. The aggregate bandwidth supported by FDDI is 100 Mbps. It uses a token-based discipline to manage multiple access. FDDI was developed with data communication in mind (rather than, for example, video or multimedia), particularly for backbone LAN interconnection in a campus or building floor-riser environment. FDDI networks are now used both as a backbone (back end) technology to connect departmental LANs, and as a front-end technology to directly connect workstations. FDDI has experienced slow acceptance because the cost of the device attachment has remained high.

Even for the traditional applications, there is a recognized potential for network bottlenecks in the near future, not only because of the increased number of users and the introduction of client/server systems (where data are not locally located at the PC but must be obtained across a distance), but also because of the increasing deployment of new graphics-intensive and/or image-intensive applications. Therefore, even systems providing an aggregate throughput of 100 Mbps, such as FDDI, may be inadequate for high-throughput applications. ATM is the scalable technology that can meet evolving requirements.

100 Mbps Ethernet systems The early 1990s saw the deployment of Fast Ethernet. The goal was to deliver 100 Mbps to the desktop cheaper than would be the case with FDDI-based adapters. The issues associated with this endeavor have been as follows: Does one keep the same CSMA/CD MAC as the 10 Mbps system, or does one move to a MAC that is more isochronous in nature? Does one aim for voice-grade unshielded wiring (type 3), use data-grade unshielded wiring (type 5), or shielded wiring (type 1)? Can one retain two-pair wiring, or will four-pair system be needed/preferred? In the opinion of a majority of interested parties, one wants to keep the existing wiring and replace the PC/workstation and hub cards for no more than a few hundred dollars. The target is to deliver 100 Mbps (aggregate shared bandwidth) to the desktop for no more than twice the cost of a 10Base-T system (which is around $250 per user). A number of proposals were made to the IEEE 802.3 committee.

In some proposals, the frame would remain the same as the Ethernet frame, but the signaling—how the frame is transmitted over the medium, as specified in the Physical and Physical Medium Dependent (PMD) sublayers—would have to be changed and then standardized. Because of these changes, new workstation/PC network interface cards and new hubs would be required (note that LLC, TCP/IP, and the applications themselves do not require any modifications). In addition, because of propagation time issues, the diameter of the network (maximum end-to-end cable length) would have to be reduced by an order of magnitude, from the current 2,500 meters to 250 meters. If the ethernets in question are interconnected over an FDDI backbone (i.e., hubs interface to an FDDI network), then the distance reduction would likely not be a problem; however, if the network hubs are connected over twisted-pair cable supporting a single logical network, then the diameter restriction will likely be a problem. A way around this problem is to utilize bridges between hubs.

The ability to operate at higher speeds without having to modify major portions of the MAC protocols has been demonstrated for some time, in both the Ethernet and token Ring context (64 Mbps Token Ring systems have been prototyped; however, new network interface cards for the PMD would be required in both cases). There has been broad vendor support for this proposal. Dual-speed (10/100 Mbps) adapters and dual-speed hub ports are envisioned. Bridging between the two systems is relatively easy (as long as the effective speed/throughput is consistent across both networks).

In other proposals the MAC protocol is also replaced, eliminating the contention scheme (which is in fact the culprit for variable delays and throughput in the network) and replacing it with a Demand Priority Protocol. Four pairs are required with this approach.

Two alternatives have emerged, in the final analysis, as standards: 100Base-T (IEEE 802.3u), and Demand Priority 100VG-AnyLAN (IEEE 802.12). Extensions have been made over the years to IEEE 802.3 Ethernet, since it was original standardization in 1983. These extensions include Ethernet switching, full-duplex Ethernet, and, most recently, Fast Ethernet (IEEE 802.3u). In particular, Ethernet switching has been fairly successful at the commercial level.*

Ethernet-based technologies (Ethernet switching, full-duplex Ethernet, and Fast Ethernet, along with hubs, routers, and adapters) tend to interwork with some degree of cohesion. Therefore, some see a continued, firmly rooted opportunity for this technology. Others see ATM as the next corporate LAN technology. It is worth noting that all the new protocols support UTP wiring, thereby eliminating a potential first hurdle related to possible deployment in the corporate landscape. However, there are significant

---

* Switching is a hub-centric form of Ethernet that takes advantage of the star topology now prevalent in LANs. It effectively eliminates Ethernet collisions by placing each workstation on its own segment (the only collisions occur when the hub and the workstation both try to transmit information simultaneously).

technology, embedded-base, internetting, and management issues impacting the actual deployment prognosis.

The IEEE 802.3u protocol maintains the 10 Mbps values of the IEEE 802.3 protocol and increases the throughput ten fold by speeding up the MAC. 100Base-T couples the ISO/IEC 8802-3 CSMA/CD MAC with a family of 100 Mbps physical layers. While the MAC can be readily scaled to higher performance levels, new physical layers standards are required for 100 Mbps operation. 100Base-T uses the existing ISO/IEC 8802-3 MAC layer interface, connected through a Media-Independent Interface (MII) layer to a Physical Layer Device (PHY) sublayer such as 100Base-T4, 100Base-TX, or 100Base-FX. With this extension, the IEEE 802.3 standard encompass several media types and techniques for signal rates from 1 Mbps to 100 Mbps [20].

It follows that the MAC procedure, MAC frame length, MAC format, and maximum number of nodes in the LAN remain unchanged. As in 10Base-T, the maximum frame size is 1,518 octets; the minimum frame size is 64 octets; and the address is 6 octets. Protocol parameter values have not changed, except for the interframe gap. The slot time for 100Base-T is 512 bit times; the interframe gap has been scaled from 9.6 microseconds to 960 nanoseconds; the transmit attempt limit is 16, while the backoff limit is 10; the jam size is 32 bits. IEEE 802.3u is designed to (inter)work with full-duplex techniques* with and without Ethernet switching. One of the features of 100Base-T is that it can operate over widely installed two-pair category 3 UTP wiring. Fast Ethernet supports three standardized signaling systems: 100Base-T4, 100Base-TX, and 100Base-FX. 100Base-TX supports two-pair category 5 UTP.

100Base-T extends the ISO/IEC 8802-3 MAC to 100 Mbps. The bit rate is faster, bit times are shorter, packet transmission times are reduced, and cable delay budgets are smaller—all in proportion to the change in bandwidth. This means that the ratio of PDU duration to network propagation delay for 100Base-T is the same as for 10Base-T. As noted, the IEEE 802.3u standard specifies a family of physical layer implementations. 100Base-T4 uses four pairs of ISO/IEC 11801 category 3, 4, or 5 balanced cable. 100Base-TX uses two pairs of category 5 balanced cable or 150 ohms shielded balanced cables defined by ISO/IEC 18801. 100Base-F uses two multimode fibers. FDDI (ISO 9314 and ANSI X3T12) physical layers are used to provide 100Base-TX and 100Base-FX physical signaling channels, which are defined under 100Base-X. For comparison

* In a full-duplex operation, one no longer has flow or congestion control. Therefore, the impact of a PDU loss is more significant than a collision because it has to be resolved at a higher layer of protocol stack (e.g., TCP). For this reason, some argue that while full-duplex operation has its place for switch-to-switch or server–to–hub connections over fiber-optic media, it is not appropriate for 100Mbps connection to the desktop.]

purposes, table 2.6 depicts wiring schemes for a number of common LANs, including 100 Mbps LANs.

Table 2.6 Wiring Schemes for a Number of Common LANs

Lan Technology	Wiring Pairs 1 pair	2 pairs	4pairs	Wiring Types Cat. 3 UTP	Cat. 4 UTP	Cat. 5 UTP	STP
IEEE 802.3; Ethernet; 10 Base-T		x		x		x	x
IEEE 802.12; Demand Priority; 100VG-AnyLAN			x	x	x	x	x
IEEE 802.3u; Fast Ethernet; 100Base-TX		x				x	x
IEEE 802.3u; Fast Ethernet; 100Base-T4			x	x	x	x	x
IEEE 802.3u; Fast Ethernet; 100Base-FX	x						
IEEE 802.3u; Fast Ethernet; 100Base-T2		x		x	x	x	x
IEEE 802.9; IsoEnet		x		x		x	
ANSI X3T9.5; TP-PMD; CDDI		x				x	x
ATM Forum; 25 Mbps ATM		x		x	x	x	x
ATM Forum; ATM CAP-4		x		x	x	x	
ATM Forum; ATM CAP-16		x		x	x	x	
ATM Forum; ATM 155 Mbps Cat. 5		x				x	x
ATM Forum; ATM 155 Mbps Cat. 3		x		x	x	x	

LAN Emulation, a first view Of the main specifications the ATM Forum is developing, perhaps the most important area concerning users in the LAN area is LAN Emulation. Traditional LANs provide a connectionless MAC service, supporting arbitration among end stations for access to a shared physical transmission medium (e.g., the twisted-pair cable). ATM, on the other hand, offers a connection-oriented communication service based on switched point-to-point physical media. To achieve connectionless MAC service over an ATM link, a protocol layer emulating the connectionless service of a LAN must be placed on top of the AAL. This layer, depicted in figure 2.6, is called the ATM MAC. This layer emulates the LAN service by creating the appearance of a virtual shared medium from an actual switched point-to-point network.

Version 1.0 of LANE was adopted in 1995 and additional work is now under way. The purpose of LANE is to provide users with a migration path to ATM without immediately incurring the high cost of implementing ATM to the desktop (however, LANE equipment must be deployed).

In legacy LANs, the membership of an individual station to a LAN segment is dictated by the physical connection of the station to the physical shared medium. Membership of a station to an ATM LAN segment is identified by logical connections to the

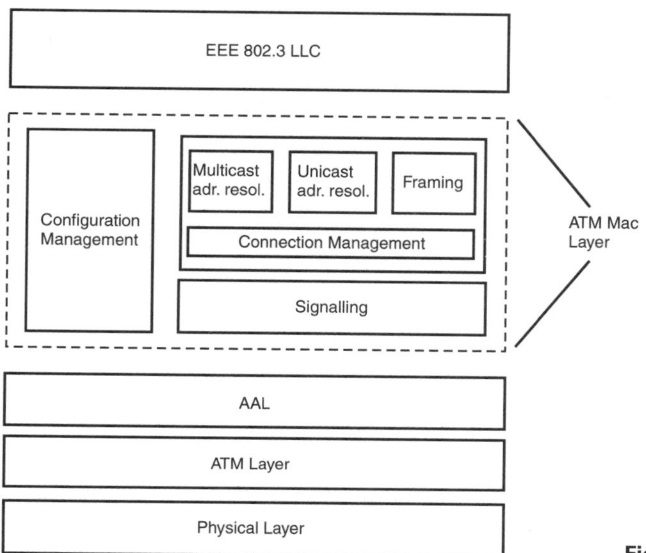

Figure 2.6 Protocol stack of LANE

multicast ATM virtual connection. Hence, membership of an ATM LAN segment is defined logically rather than physically; the membership information is stored in some management database. This capability of ATM LANs offers terminal portability and mobility. LANE does provide transparent support for LAN- based applications, since it functions as a bridge at layer 2. LANE is a converting-bridge technology between the connectionless Ethernet/Token Ring environment and the connection-oriented ATM environment. It also supports ATM-enabled devices to communicate with LAN Emulated devices. LAN Emulation allows for logically separate emulated LANs to coexist on the same physical ATM network.

LAN Emulation does not allow users to leverage the end-to-end class of service functionality that ATM provides in end-systems; however, it will provide for a higher bandwidth and a more stable network infrastructure for large building and campus backbones. It also requires that Ethernet, Token Ring, and FDDI be separate, since it does not provide transparent bridge functions between these technologies.

2.3.2 Wide area connectivity

There are several factors characterizing wide area services that are of particular relevance to client/server (and intranet) environments. Table 2.7 highlights some of these characteristics; however, not every possible combination shown in this table corresponds to a service that can be secured from a carrier. Connection-oriented communication highlighted in the table is similar to the traditional dedicated-line or circuit-switched

Table 2.7 Key Factors for Wide Area Connectivity

Characteristic	Typical Ranges
Bearer Mode	Connection-oriented or connectionless
Connection Type Supported (Signaling Support)	Point-to-point, point-to-multipoint, multipoint-to-multipoint
Geographic Scope of Coverage	IntraLATA, interLATA, international
Speed (bandwidth)	Nx64 kbps, T1/DS1/E1 (1.544–2.048 Mbps), T2/DS2 (6.312 Mbps), T3/DS3 (44.736 Mbps), STS-1 (51.84 Mbps), STS-3c (155.250 Mbps), STS-12c (622.08 Mbps)
Switching type	None (dedicated line), circuit switched, packet-switched (frame relay, cell relay)
Symmetry	Bidirectional symmetric bandwidth, bidirectional asymmetric bandwidth, unidirectional

environment where the user goes through a connectivity setup phase, an information transfer phase, and a connectivity teardown phase (for a circuit-switched call, this is done in realtime; for a dedicated-line service, the setup is done at service initiation time and the teardown is done at service cancellation time). In a connectionless service, each data unit is treated independently, and no connectivity setup/teardown is required. This is similar to the way an information frame is transferred in a LAN.

Strategically, higher-speed WANs will be required; however, in the short term any number of technologies will probably be adequate to various degrees. Table 2.8 depicts some of the key telecommunication services now becoming available in support of distributed computing. Table 2.9 groups these services into four types: dedicated/switched

Table 2.8 High-Speed Service and Related Terms

Cell Relay Service	This is a method to multiplex, switch, and transport a variety of high-capacity signals, using short, fixed-length data units known as cells. The Asynchronous Transfer Mode is the accepted international standard for the cell structure. It uses a 155- or 622-Mbps public switched WAN service (service is also possible over a private switch), where user's cells are delivered at high speed to a remote destination (or destinations). It is both a Permanent Virtual Connection (PVC) service and a Switched Virtual Connection (SVC) service.
Fractional T1	This is a point-to-point dedicated service supporting Nx64 kbps connectivity (typically, N=2, 4, 6, 12).
Frame Relay, Private	This is a multiplexed service obtained over a private high-speed backbone equipped with appropriate nodal processors (fast packet switches). Used to interconnect LANs at Nx64 Kbps or 1.544 Mbps.
Frame Relay, Public	This is a multiplexed service provided by a carrier. The user has a single access line into the network and can deliver frames to remote users without having to provide dedicated communication links or switches. Used to interconnect LANs at Nx64 Kbps or 1.544 Mbps.

Table 2.8 High-Speed Service and Related Terms

ISDN H0	This is a switched service providing physical connectivity at 384 Kbps, using the ISDN call setup mechanism.
ISDN H11	This is a switched service providing physical connectivity at 1.536 Kbps, using the ISDN call setup mechanism.
SMDS	Switched Multimegabyte Data Service is a public switched service supporting connectionless cell-based communication at 1.544 and 45 Mbps access speed, targeted for LAN interconnection.
SONET	Synchronous Optical Network (SONET— known as Synchronous Digital Hierarchy outside the U.S.), is a specification for digital hierarchy levels at multiples of 52 Mbps. It is also a point-to-point dedicated service supporting Nx52 Mbps connectivity over fiber-based facilities.
Switched T1	This is a switched service providing physical connectivity at 1.536 Mbps.
T1 (DS1)	This is a point-to-point dedicated service supporting 1.544 Mbps aggregate connectivity.
T3 (DS3)	This is a point-to-point dedicated service supporting 44.736 Mbps aggregate connectivity or 28 DS1 subchannels.

Table 2.9 Classification of Key WAN Services in Support of Client/Server Systems

	Nonswitched	Switched
Low Speed	Analog private line DDS private line Fractional T1 private line T1 private line Frame relay (permanent virtual circuit)	Dial–up with modem ISDN Packet–switched network Frame relay switched virtual circuit
High Speed	T3 private line SONET private line ATM/cell relay service (permanent virtual circuit)	Switched Multimegabit Data Service (SMDS) ATM/cell relay service (switched virtual circuit)

and low speed/high speed. Figure 2.7 shows the applicability of some of the key interoffice/long-distance high-speed services plotted against the burstiness requirement of the application (burstiness is the ratio of the instantaneous traffic to the average traffic) [2]. Figure 2.8 shows an example of a client/server system implemented across a WAN.

Cell relay service Asynchronous Transfer Mode is a high-bandwidth low-delay switching and multiplexing communication technology supporting wide-area communications; as noted, the same technology is now also being applied to the development of next-generation LANs (to be exact, ATM refers to the network platform, while cell relay service refers to the actual service obtainable over an ATM platform). It is the general industry consensus that ATM is the WAN service of choice for applications requiring high throughput. It is a connection-oriented technology supporting a number of

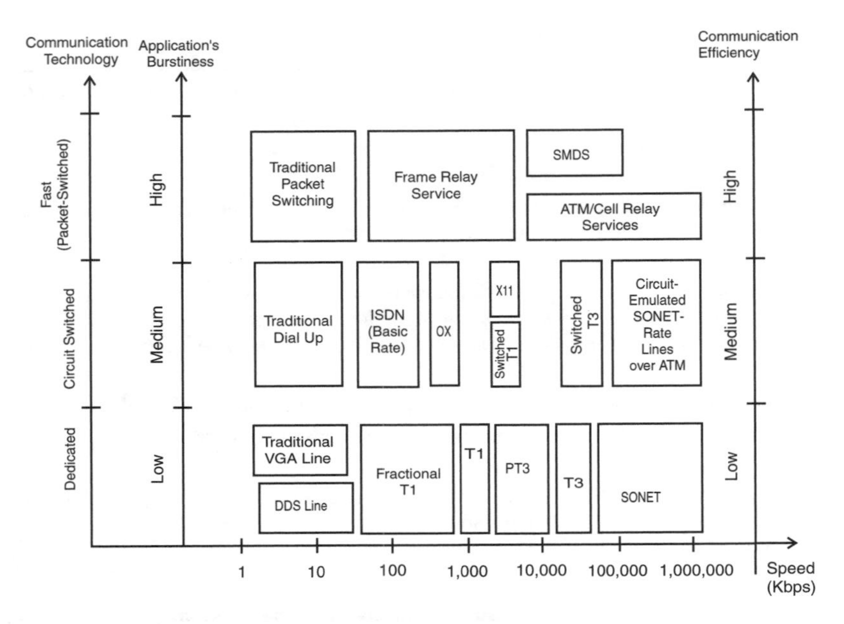

Figure 2.7 Key WAN services available commercially

categories of statistical multiplexing. For readers familiar with the operation of a protocol stack it is simply a matter of realizing what functional partitioning has been instituted by the designers of ATM, and what the peer entities in the user's equipment and in the network are. The cell relay protocols approximately equate to the functionality of the MAC/LLC layers of a traditional LAN, but with the following differences: random access is not utilized, channel sharing is done differently, and the underlying medium may be different [21].

Two remotely located user devices (say a client and a server) needing to communicate over an ATM network can establish one or more bidirectional virtual (i.e., not hard-wired and/or dedicated) connections between them to transmit cells (fixed-length packets 53 bytes long). This connection is identified by each user by an appropriate identifier, similar in some respects to the way virtual channels are identified in a packet-switched network. Once such a basic connection is set up, user devices can utilize the virtual connection-oriented channel for specific communication tasks. Each active channel has an associated bandwidth negotiated with the network at connection setup time. The transfer capacity at the User Network Interface (UNI) is 155.52 Mbps at a 44.736 Mbps rate.

Connections in an ATM network support both circuit-mode and packet-mode services of a single medium and/or mixedmedia and multimedia. ATM carries several types

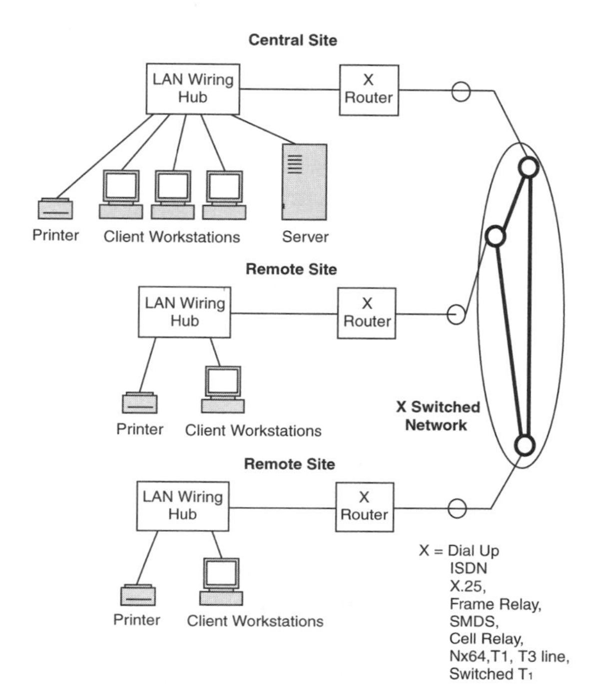

Figure 2.8 Client/server system across a WAN

of traffic, including Constant Bit Rate (CBR) and Variable Bit Rate (VBR)— for example, video transmission generates CBR traffic, while data applications (say, router traffic for a traditional LAN) generate VBR traffic. CBR transfer rate parameters for on-demand services are negotiated at call setup time (CBR transfer rate parameters for permanent services are agreed upon with the carrier from which the user obtains service). VBR services are described by a number of traffic-related parameters (peak rate, sustained rate, maximum burst length, etc., which are also negotiated at call setup time). For wide area communication, cell relay entails the following:

1. Specification of the network interface configuration—that is, the required protocols and procedures the user needs to implement in his or her equipment in order to obtain the high-speed, high-quality cell relay service. The user equipment can be comprised of a variety of elements (workstations, routers, servers, multiplexers, etc.). The access speed and other QoS factors can be specified by the user. The interface covers both the information flow and the call control flow (i.e., the user has a *transmit* channel and a *signaling* channel).
2. A high-speed, typically fiber-based, local loop to enable the user's information stream originating at a user's location to reach the broadband switch at the serving central office or some other hub location, where it is appropriately handled.

3. A high-speed broadband switch which is able to interconnect users at the required bandwidth rates.
4. A call control capability (at the broadband switch or other location) to accept the users' service requests and to allocate network resources to satisfy these requests. The call control capability supports point-to-point, point-to-multipoint, and multipoint-to-multipoint connectivity.
5. A network capability to maintain the requested bandwidth, connectivity, and QoS.
6. An interoffice high-speed network infrastructure supporting wide area connectivity.
7. An interoffice (overlay) signaling network to carry the users' service requests (i.e., signaling information) to the appropriate destination.

The access protocol in the user equipment consists of a physical layer at the lowest level and an ATM layer over which it provides information transfer for all services. Above the ATM layer, the ATM Adaptation Layer (AAL) provides service-dependent functions to the layer above the AAL (these layers above the AAL are somewhat similar to TCP/IP in a traditional LAN; in fact, TCP/IP may continue to be used by users' PCs and hosts— note that AALs usually only go as high as the data link layer). The AAL protocols are implemented in the users' equipment. The Service Data Units reaching the AAL consist of user information coming down the protocol stack (e.g., from a TCP/IP stack or from a video codec); the information is segmented/cellularized by AAL into 53-octet cells, so that it can be efficiently shipped through the network. The AAL enhances the services provided by the ATM layer to support the functions required by the next higher layer. The AAL-specific information is nested in the information field of the ATM cell.

Wide area networks providing cell relay services started to appear in 1994, with more widespread deployment soon thereafter. There is keen interest in ATM and cell relay service on the part of local exchange carriers, interexchange carriers, and international carriers. Cell relay service provides both a Permanent Virtual Connection (PVC) and a Switched Virtual Connection (SVC) service. A PVC implementation establishes a fixed path through the network for each source-destination pair, which remains defined for a long period of time (weeks, months, or years). In SVC, resources are put in place only for the duration of the actual session (minutes or hours). Cell relay service is expected to be used in large companies.

This topic will be treated in more detail in chapter 6.

Frame relay service Frame Relay Service (FRS) is a connection-oriented service operating at Nx64 Kbps or 1.544 Mbps (2.048 Mbps in Europe). It is offered by a variety of carriers, including the Regional Bell Operating Companies, and interexchange carriers. It started to be available in early 1991 and by press time it was available in most U.S. cities and in many European cities. FRS is positioned for LAN interconnection. Given the relatively low speed, it is marginally useful for high-speed applications. Compared to X.25-based service, FRS aims at reducing network delays, providing more efficient bandwidth utilization, and decreasing communication equipment cost. The increase in efficiency and reduction in delay is achieved by performing error correction on an end-to-end basis rather than on a link-by-link basis, as is the case, for example, in traditional packet switching. This makes the protocol much simpler. Communication links are now carried in an increasing number on fiber-optic facilities, making them cleaner, as measured by the Bit Error Rate (BER). Because the circuits are much cleaner, it is more effective to perform error management on an end-to-end basis. Most of the applications to date have been for wide area LAN interconnection, where the LANs support traditional data-only applications. Some experimentation in support of video conferencing has been reported, but with limited success.

A frame relay interface can support multiple sessions over a single physical access line. Equipment implementing frame relay interfaces have been implemented on products such as LAN bridges and routers, as well as on T1 multiplexers. Frame relay provides both a PVC and SVC service. Frame relay service is expected to be used in medium to large companies.

Digital dedicated line services In spite of the emergence of other digital services, high-speed dedicated digital lines operating at Nx64 Kbps (N=1, 2, 6, 12), 1.544 Mbps (also known as T1) and 45 Mbps (also known as T3) still represent a common way to interconnect remote LANs. A dedicated line implies that the entire bandwidth can be applied to the interconnection task (unless a portion of the bandwidth is allocated to another application). Since the bandwidth is not generally shared, there is no delay variation, and there is no frame discard over the WAN, dedicated lines are somewhat better suited for high-performance applications, compared with FRS. This approach, however, does have at least three drawbacks:

1. Relatively high cost
2. Multitude of lines, growing at 0.5 times the square of the number of locations to be connected, implying high communication and network management cost
3. Inflexibility to reach *off-net locations*, typically providing connectivity between only two or a few sites.

High-speed dedicated services are expected to be used in medium to large companies.

Other technologies For a thorough description of the other technologies, the reader may reference [22].

2.4 References

1. D. Minoli, A. Schmidt, *Client/Server over ATM,* Greenwich CT, Manning Publications, 1997.
2. T. E. Bell, "Jobs at Risk," *IEEE Spectrum*, August 1993, pp. 26.
3. R. Moskowitz, "What Are Clients and Servers Anyway?" *Network Computing, Client/Server Supplement*, May 1993.
4. Data Communication, *Sharing the Load: Client/Server Computing,* March 21, 1989, pp. 19-29.
5. J. C. Panettieri, "How to Break Through the Logjam," *Network Computing, Client/Server Supplement*, May 1993.
6. L. Berg, *Implementing Client/Server Computing,* paper presented at COMNET 92, Washington, DC, Jan. 1992.
7. K. Myhre, *Please Explain Client/Server,* paper presented at COMNET 92, Washington, DC, Jan. 1992.
8. D. Ferris, "Client/Server Database Models Are Emerging," *Network World*, May 11, 1992.
9. S. Morse, "Client/Servers Is Not Always the Best Solution," *Network Computing, Client/Server Supplement*, May 1993.
10. D. Minoli, *First, Second, and Next Generation LANs*, McGraw-Hill, New York, 1993.
11. Promotional Material, *Network Computing, Client/Server Supplement*, May 1993.
12. P. Smith, *Client/Server Computing,* Sams Publishing, Carmel, Ind, 1992.
13. D. Minoli, *Imaging in Corporate Environments,* McGraw-Hill, New York, 1994.
14. D. Minoli, *Third Generation LANs,* Proceedings of TEXPO 1993, San Francisco, April 6-8, 1993.
15. J. Cox, "Oracle to Cast SQL*Net at Distributed Apps," *Communications Week*, August 2, 1993, 1 ff.
16. J. Cox, "Users Urge: See Client/Server Clearly," *Communications Week*, June 21, 1993, 11 ff.
17. T. Wilson, "IBM Continues Client/Server Shift," *Communications Week*, June 21, 1993, pp. 4.
18. D. Minoli, "ATM Makes Its Entrance," *WAN Connections—Network Computing Magazine*, August 1993, 22 ff.

19 D. Minoli, *Analyzing Outsourcing,* McGraw-Hill, New York, (1994)

20 A. Freedman, *Electronic Computer Glossary,* City, The Computer Language Co., 1993

21 H. Newton, *Newton's Telecom Dictionary,* (Miller Frieman Books, 1998)

22 D. Minoli, *Enterprise Networking, from Fractional T1 to SONET, from Frame Relay to BISDN,* Artech House, Norwood, MA, 1993

23 D. Minoli and A. Alles, *LAN, ATM, and LAN Emulation Technologies,* Artech House, Norwood, MA, 1997

CHAPTER 3

Key Internet and Intranet Protocols

This chapter describes, in summary form, the basic middle-layer communication protocols that form the foundation for intranets. These protocols are TCP, UDP, IP, and PPP, which have evolved from Internet applications. Lower-layer protocols and facilities, specifically ATM, are described in Parts II and III. Higher-layer protocols, including HyperText Transfer Protocol (HTTP), are described in chapters 4, 5, and 6.

3.1 Introduction to the Internet

The Internet is a federation of computer networks utilizing the same communication protocols [1]. The subnetworks that comprise the Internet are connected via high-speed links or other modern communication services. The Internet is an informatics cooperative. It is not owned by anyone. There are three categories of entities in the Internet community:

1. Information providers
2. Users
3. Connection providers

Information providers make information available to users, the people who access the data. Connection providers provide the network connection for both information providers and users.

The Internet started, in its original form, in the late 1960s as the Advanced Research Projects Agency (ARPA) experiment in packet-switched networking. The network, known as ARPANET, was created by the Department of Defense (DoD) in order to provide researchers with remote access to computers. The progenitor of the Internet came to light in 1969, when a group of DoD researchers linked four computers at UCLA, Stanford Research Institute, the University of Utah, and the University of California at Santa Barbara to establish a network to communicate with one another, in support of government projects. Three years later, more than 50 universities and military agencies were linked via ARPANET, and other computer networks began to appear around the country and around the world.

In the early 1980s, the original ARPANET was separated into two networks: the new ARPANET, which was still an experimental network, and MILNET, which was a military network. Connections existed between the two networks to allow for communication across the communities. The new network was named DARPA Internet and later, the Internet. In the late 1980s, the National Science Foundation (NSF) established a new network, the NSFNET, to allow communication between its supercomputer centers. In 1990, with many of its members connected to newer, higher-speed networks, the

ARPANET ceased to exist [2, 3]. Today, thousand of subnetworks, some from the lineage just described, and others totally new, form what we know as the Internet.

The basic technology used at the beginning was packet switching, a method of fragmenting messages into smaller parts called *packets*, routing them to their destinations, and reassembling them. The concept of packet switching had been developed, at the research level, in the mid-1960s. The concept of segmenting data (either at the data link layer or at the network layer) has remained with us in the constructs of Ethernet frames, frame relay frames, ATM cells, and packets.

There are several advantages to segmenting data. Segmentation facilitates sharing of the same connection by multiple users by fragmenting the information into discrete units, which can be routed over multiple distinct virtual circuits in the same physical circuit. Because of the potential for transmission errors, especially prior to the widespread deployment of fiber-based facilities, packet switching allowed corrupted packets to be retransmitted without requiring that the entire message be sent again. Packets may carry information about themselves, which network nodes they have traversed, and where they are going. Packetization allows effective backbone bandwidth sharing, particularly in bursty data environments, where one user does not have enough real-time transmission requirements to fully load a link [4-21]. The ARPANET utilized Network Control Protocol as its packet transmission protocol until the early 1980s, when it was replaced with the now well-known TCP/IP protocol.

The vision of a "network of networks," had its origin in 1972 at the First International Conference on Computer Communications. Participants from network computing communities around the world discussed the need for protocols to enable the interconnection of dispersed subnetworks.

As the ARPANET evolved, the military and educational networks took different paths. Non-military government agencies, such as NASA, also soon began to experiment with large computer networks. Within a short time these various networks began to interconnect. From 1985 to 1986, the NSF initiated a project to improve the quality of scientific research in the U.S. by establishing a data network linking major supercomputer centers and computer networks, using the TCP/IP protocol suite. TCP/IP was chosen because it was (and is) a non-proprietary protocol, was used on the ARPANET, and was favored by users in the scientific and DoD communities. As a result of this project, a gamut of regional and state computer networks came into existence. Subsequently, the NSF outlined a plan to interconnect them. In 1986, the NSF created the NSFNET backbone, a national network of 56 Kbps lines linking the NSF-funded supercomputer centers. This backbone network also served to link the regional networks via the regional connections to supercomputer centers. Initially these connections were achieved by running a connection from the center to the campus or to the research center. This soon became too expensive, and NSF begun to run connections to the nearest site that was already connected [4-21].

In a few short seasons, the NSF reached the conclusion that the infrastructure of the network needed to be upgraded, and they did not want to be in the network administration business. IBM, MCI, Merit, and the state of Michigan won a bid to upgrade and manage the new network. The agreement was for five years, with an option for renewal. In this arrangement, MCI would provide the DS1 lines, IBM would provide routers, and Merit would manage the network. By 1988, this network was operational and connected 16 sites. In 1990, IBM, MCI, and Merit established Advanced Network and Services, Inc. (ANS). ANS furnished and operated the circuits and routers of the NSFNET backbone and served in a technical capacity known as the *Internet Routing Authority* [4-21].

In the early 1990s, network traffic started to increase as more sites were coming online, facilitated by simpler and higher-speed access. It was now possible to transmit large quantities of data over the network in a short time. This increased the growth rate, so that the total network traffic doubled faster than once a year. NSF realized once again that it was going to need a major upgrade to its backbone. ANS proposed upgrading the backbone to DS3 lines, and NSF agreed. By the end of 1991, all the NSF sites were connected by this new backbone, but the full migration took three years. There were now 4,500 different networks connected to the NSF backbone. Also, the number of packets traveling the network had increased from 160 million per month in 1986 to 12 billion per month in the early 1990s. The years 1991 through 1994 were years of significant growth. Many new networks were added to the NSF backbone. Hundreds of thousands of new hosts were added. Also, significant new uses of the Internet were developed, as the World Wide Web WWW quickly became the second most used application on the Internet, based on the amount of data transferred [4-21].

As the NSFNET grew, the term *Internet* came to refer to the totality of interconnected networks using the TCP/IP protocols. These included not only the NSFNET backbone and regionals, but the government networks and various foreign networks. The Internet now connects sites in about 70 countries; it connects over 25,000 networks with over 2,400,000 computers on them and is growing at a monthly rate of about 10 percent. In addition, gateways to commercial networks, such as MCIMail, Prodigy, and CompuServe, allow for electronic message exchange with over 130 countries all over the world [4-21].

In 1993, the NSF proposed a new architecture, comprised of a very high-speed Backbone Network Service (vBNS) (which operates at a speed of 155 Mbps), Network Access Points (NAPs), a Routing Arbiter, and many Network Service Providers (NSPs) to carry national traffic. The motivation was to increase competition: although the

Merit/ANS/IBM/MCI partnership had been successful, it did not represent, in the view of many, an open and fair environment for networking services [4-21]. In 1995, an important change was announced in the way the Internet operated. NSF stated that as of May 1, 1995, it would no longer allow direct access to the NSF backbone. Merit/ANS/IBM/MCI terminated the NSFNET backbone service.

The new NSF directly funds the Routing Arbiter and the vBNS, but use of the vBNS (operated by MCI) is restricted to organizations that require high speeds for applications such as scientific computation and visualization. Merit is continuing its work in national and international networking as the Routing Arbiter, under a five-year award from NSF that began in July 1994 [4-21]. NSPs are not funded directly by NSF. Instead, NSF (temporarily) funds regional network attachments to NSPs, with support declining to zero funds over four years. NAPs interconnect the vBNS and other backbone networks. The NSF contracted with four companies to be NAPs to the NSF backbone. They are operated by PacBell in San Francisco, Ameritech in Chicago, Sprint in New York, and Metropolitan Fiber Systems (MFS) in Washington, D.C. These companies can now sell connections to groups, organizations, and other companies. This new arrangement was undertaken for two reasons:

1. The U.S. government wanted to exit the communications business.
2. A desire for the commercialization of the Internet: when the network was operated by the U.S. government, there had been restrictions on its use. Companies could only use links to the NSF backbone (sponsored with government funds) for research. They could not use it for advertising or other commercial activities. Companies had to be careful not to send out product information over a communication line that had been installed with a grant from NSF.

The shift created an incentive for three major Internet access providers: Performance Systems International, Inc. (PSI); UUnet Technologies/MFS; and General Atomics Cerfnet, to create their own commercial backbones*. These providers, along with nine others, formed the Commercial Internet Exchange (CIX) [4-21].

---

* Key players include ANS, Sprint, MCI, UUnet/MFS/PSI, and AGIS.

3.2 Communication protocols originating in the Internet

The concept of interconnected networks has clearly originated in the Internet. The same technology has since been deployed to the enterprise network and the intranet. This section provides a brief description of the key protocols in the TCP/IP suite.

TCP/IP is the protocol that supports transmission of all information moving across the Internet. The TCP/IP protocol suite consists of multiple protocols. TCP is the transport layer above IP that ensures reliable connections between sender and receiver. TCP provides a virtual connection, so that two hosts on one or more networks can communicate, independent of the technology used in each. IP, the Internet Protocol introduced in chapter 2, is a network-layer protocol, which moves data between host computers within a subnetwork. IP defines the unique addresses for the network and its hosts. The IP protocol supports two key functions [3, 22]:

1. Packet routing through the different networks—an IP router can transmit a packet to either the receiver or to the next network gateway (router).
2. Fragmentation and assembling of packets that are too long for an intermediary subnetwork.

UDP (User Datagram Protocol) is another protocol in the TCP/IP suite; it is less complex than TCP, but it is also less reliable. It tends to be used for short, atomic transactions. Another protocol in the TCP/IP suite is Internet Control Message Protocol (ICMP). ICMP carries network error messages and reports conditions that require attention by network software [4-21].

Above TCP/IP, application-based protocols provide higher level services. File Transfer Protocol (FTP) allows file transfers between two network devices. FTP servers provide collections of documents, programs, images, and stored multimedia information. To overcome the cost of long-distance transmission, many popular FTP servers are geographically duplicated or mirrored (Archie servers are databases that index documents from different FTP servers, which facilitates locating a specific file). Network News Transfer Protocol (NNTP) is the bulletin board protocol. NNTP allows an Internet host to participate in any of thousands of forums. Simple Mail Transfer Protocol (SMTP) is the protocol permitting electronic mail transfers between Internet hosts.

3.2.1 IP addressing

An Internet address is an IP address. IP addresses can normally be associated with a host computer, although some computers on the Internet have multiple addresses. IP addresses consist of four octets, whose values are represented decimally (separated by periods) for example: 136.102.231.59. This is called dotted-decimal notation. The first portion provides network addressing and the last portion provides host addressing; the point of separation depends on the address class. The Internet Network Information Center (InterNIC) Registrar is responsible for the assignment of valid network addresses. The InterNIC ensures that each network has a unique network identifier.

The hierarchy is as following: geographical, organizational, departmental, and server. Name servers allow for the communication of logical addresses rather than physical addresses. Figure 3.1 is a model of Internet connectivity. [3,24]

An IP address is comprised of a network address (netID) used to identify the network to which a device (also called a host) such as a PC, terminal, or computer, is connected and an identifier for the device itself (hostID). An IP address can be represented as:

```
AdrType|netID|hostID.
```

Traditionally, there have been four IP address classes on the Internet (see figure 3.2).

Class A addresses start with a 0 in the first bit and use the next seven bits for the network number and the remaining three bytes for the hostID. There are only 127 networks

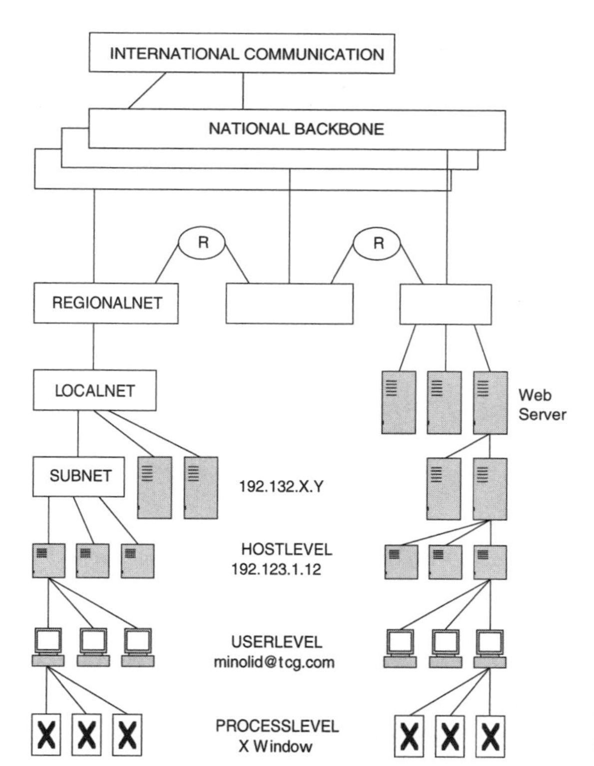

Figure 3.1 Model of Internet interconnectivity

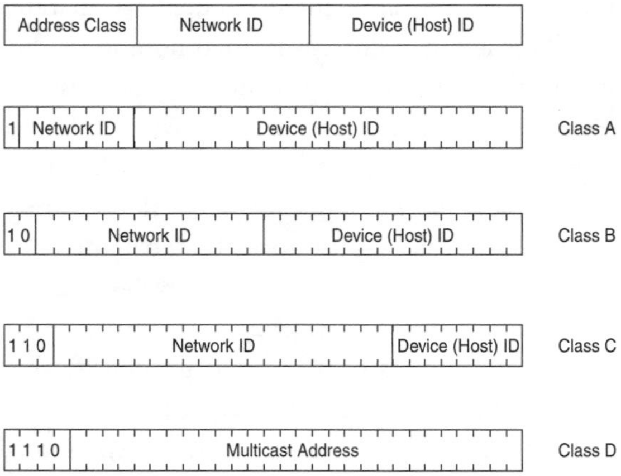

Address Class | Network ID | Device (Host) ID

1 | Network ID | Device (Host) ID — Class A

1 0 | Network ID | Device (Host) ID — Class B

1 1 0 | Network ID | Device (Host) ID — Class C

1 1 1 0 | Multicast Address — Class D

Figure 3.2 Structure of IP address

with Class A addressing. Class A networks can attach up to 16,777,216 hosts. Because of this, Class A addresses are reserved for networks with greater than 65,536 hosts.

Class B addresses start with a 10 in the first two bits and use the next 14 bits for the network number and the remaining two bytes for the hostID. Thus, the Internet can have up to 16,384 Class B networks, which can attach up to 65,536 hosts each. Typically, networks with more than 256 host computers are given Class B addresses.

Class C addresses start with 110 in the first three bits and use the next 21 bits for the network number and the last byte for the hostID. The Internet can have 2,097,152 individual networks with Class C addresses, but each of these networks is limited to no more than 256 host computers.

Class D addresses (start with 1110) are used for multicast addresses, which represent a group of Internet host computers. Multicasting delivers messages to one or more host computers.

IP addressing also allows a portion of the host/device field to be used to specify a private subnetwork (the netID portion cannot be changed). Subnetworks are an extension to this scheme by considering a part of the `<host address>` to be a *local network address*—that is, a subnetwork address. IP addresses are then interpreted as:

`<network address>< subnetwork address><host address>.`

In Class A addressing for example, a subset of the bits from bit 8 to bit 31 could be employed for subnetwork identification. The partition of the original `<host address>` into a `<subnetwork address>` and `<host address>` part can be done by the local administrator without restriction. However, once this partition has been established, it must be used consistently throughout the whole local network. Also, whereas bits can in theory be used freely, it is best to employ a contiguous set to represent the subnetwork.

The address space is quickly being used up on the Internet. While a new version of IP (IPv6) has been worked out that supports billions of addresses, it is not yet widely deployed; in any event, existing addressing (IPv4) will continue to be supported for a time [4-21].

3.2.2 Domain Name Servers

In 1983, the University of Wisconsin developed the idea of the Domain Name Server (DNS). This meant that the sender of a message no longer needed to know the exact path (the IP number) to the destination. To ensure successful Internet communication, a fast and reliable method of translating between address schemes must be available. A DNS is a program that translates domain names into IP addresses. Thousands of computers on the Internet contain name-server software.

If one wanted to send a message to `resource.cso.uiuc.edu`, the message would first go to the DNS that served `edu`, which would then tell it how to get to `uiuc`. This would tell the message here it could find `resource.cso`. This was a significant improvement over the old system, where the sender either had to send a message to the IP number, or to the administrator of the system that the user was using. The administrator in turn would have to place an entry in the host's file that would transform the name into an IP number. Thus, DNS naming made administration of machines much easier and also made life easier for the users, since people find it easier to remember names instead of numbers [4-21].

The Internet uses a hierarchical naming scheme, which decentralizes authority and responsibility for assigning names. The top node is not named, but each domain under the root has a name, as do the levels under them. The level immediately under the root consists of several domain types.

Arpa—special Internet domain, which maps dotted-decimal IP addresses to domain names
com—commercial organizations
edu—educational organizations
gov—U.S. government organizations
int—international organizations
mil—U.S. military organizations
net—a network that does not fit one of the above
org—an organization that does not fit one of the above

When a user's application is about to connect to a particular Internet host, it will first contact a name-server program and request a DNS lookup service, which will usually return the 32-bit IP address. Many organizations on the Internet dedicate a computer to the name server task. Name servers are hierarchical in that they do not need to know each IP address on the Internet. Name servers can, however, resolve any address on the Internet by communicating to other name servers [4-21].

3.2.3 TCP/IP protocol suite: an overview

Figure 3.3 shows how data flow through the various protocol layers associated with the TCP/IP stack. The lowest level of the protocol stack is known as the physical layer and consists of the network transmission interfaces to the transmission line. The top layer is the application layer; it represents the interface to the client/server program. As data move through the protocol stack, they flow from the application layer down to the physical layer and across the network. When data arrive at the physical destination, they flow up though the protocol stack towards the destination application. As data flow through the protocol stack, the TCP/IP protocols may partition these data into smaller segments [4–21].

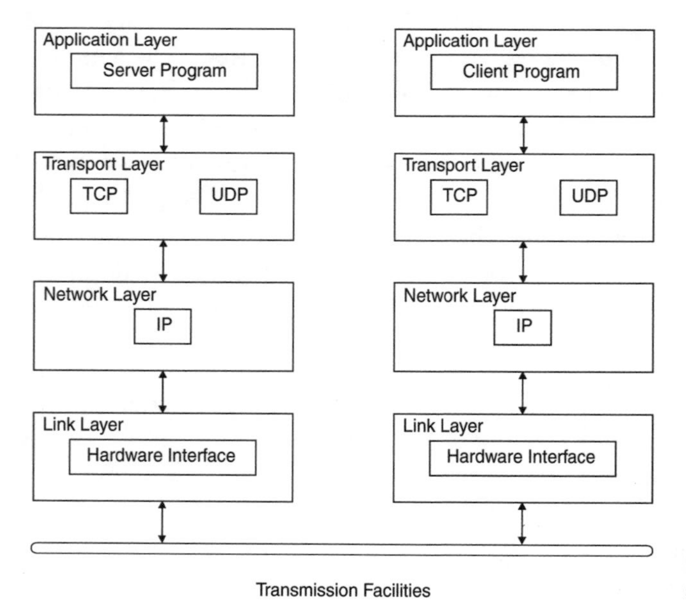

Figure 3.3 TCP/IP protocol suite

TCP is a connection-oriented protocol. It must establish a virtual connection with the receiving peer before actual communication can occur. TCP needs this virtual connection, because it has to transfer its data in sequence. TCP is a reliable protocol in that it utilizes checksums (an error-checking method of adding appropriate bits to each data block) to verify that the data received have not been corrupted during transit. Upon completing the checksum, TCP acknowledges each message (individually or as part of a block) to verify that it has been received properly. An acknowledgment is sent to the sending peer by the receiving peer for each message block or group of message-blocks successfully received [4–21].

IP is a connectionless protocol: it does not have to establish a connection before transmitting information. Each IP message must carry delivery address information. IP is an unreliable protocol in that it does not require acknowledging of received messages. IP is also a datagram protocol, because it transmits information as self-contained units of information traditionally called "datagrams" (IP-level PDUs). The IP protocol does not require a virtual circuit because the sequence of datagrams sent does not have to get to the destination in order. Thus, datagrams can be sent over different paths across the network. Resequencing must occur outside IP.

UDP, like IP, is a connectionless protocol; it, however, operates at the transport layer (the same layer as TCP). UDP is considered to be an unreliable protocol, because it does not utilize checksums or acknowledgments. UDP also uses a datagram protocol, specifically IP. UDP does not require a virtual circuit to be set up before it begins transmission of data [4–21].

The following sections provide a more detailed description of the key protocols in the TCP/IP suite.

3.2.4 Internet protocol

IP is the delivery system for the TCP/IP protocol suite. All the other protocols in the suite (i.e., TCP, UDP, and ICMP) use IP for data delivery. Hence, an IP datagram encapsulates the data from the other protocols in the TCP/IP suite. As data pass through each layer in the protocol stack, they are encapsulated for the next lower level—that is, they are packaged in the format required for the next layer. Each layer builds on the previous layer's encapsulation. Because IP is a connectionless, unreliable protocol, each IP packet sent must contain an IP header in addition to the data being sent. This header contains required information. See figure 3.4.

The header includes the version number of the IP protocol being used to allow IP software to reject incompatible packets, should there be any. It contains the header length, which allows the protocol to know where the payload data begin. The type of service (TOS) defines the priorities of the datagram and allows the network layer to make decisions on the datagram based on its priority. IP headers include a field containing the total length of the datagram (including the data). IP datagrams may be fragmented as they cross the internet; that is, if a datagram reaches a router that handles a smaller datagram size, then it will be fragmented into smaller datagrams so the router can process it. When fragmentation occurs, the total length of the IP datagram is very

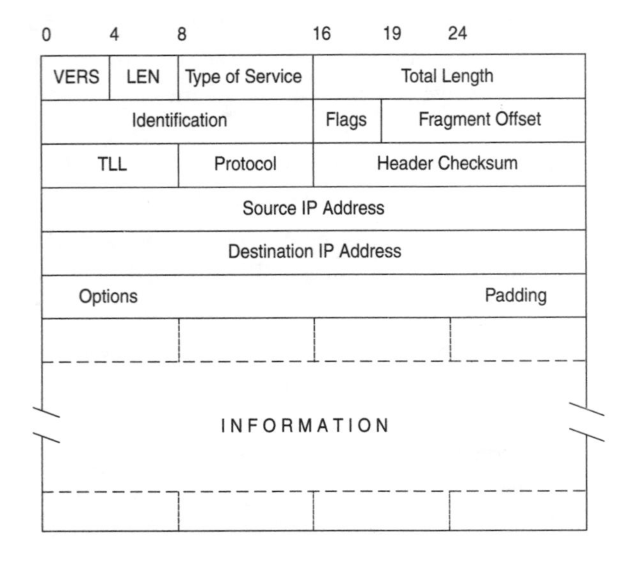

Figure 3.4 Protocol data unit

important for the router in computing new datagrams [4–21]. The header also contains a flag and fragment offset field, which are used in conjunction with the identification field to reassemble fragmented datagrams.

The identification field is used to uniquely identify each datagram (or message) so that, the receiving application will know which IP datagrams belong to which message. The time-to-live (TTL) field tells how long a datagram can live on the Internet. If a datagram gets lost, the TTL field will allow routers on the Internet to check how long the datagram has existed and delete the datagram if it has been around for too long. The checksum field will detect if the data have been corrupted during transit and the protocol field tells what kind of data (TCP, UDP, ICMP) has been encapsulated in the IP datagram. The source and destination fields contain the 32-bit IP address of the sending host and the receiving host. Finally, the IP options provide features designed to let network professionals test and debug network applications [4–21].

Fundamental to delivering IP datagrams is the IP routing table, which stores addresses for selected destinations on the network. Network software can search a routing table to find the best way to reach a specific destination. Each network interface on the same physical network must have the same netID number, but it must have a unique hostID number. IP uses a routing table to route packets between networks, not host computers, routing tables use netID numbers only [4–21]. Each routing table contains three fields: network, gateway, and flags. The first two fields contain netID numbers. The flags field identifies networks that directly connect to the owner of the routing table. The network field contains a list of netID numbers. The gateway field is a router field identifying a router on a path that leads to the network identified in the network field. Routing tables only contain the next hop in the path to a particular destination, not the entire path. Table 3.1 shows a sample of a routing table for a router. The table shows that router B is directly connected to networks 100.0.0.0 and 200.0.0.0. It also shows that networks 130.0.0.0 and 140.0.0.0 are indirectly connected through 100.0.0.2; thus, 100.0.0.2 is the next hop on the path to networks 130.0.0.0 and 140.0.0.0.

Table 3.1 A Typical Routing Table

Network Destinations	Route
100.0.0.0	Direct delivery
120.0.0.0	Direct delivery
130.0.0.0	100.0.0.2
140.0.0.0	100.0.0.2

For an extensive description of how these routing tables are maintained and updated, see reference [25].

Two protocols related to IP are the Address Resolution Protocol (ARP) and the Reverse Address Resolution Protocol (RARP). ARP maps IP addresses in the network layer to the corresponding address in the link layer. ARP mapping is dynamic, since it automatically remaps addresses when the network layer configuration changes. For example, a router may need to know the MAC address of a station on its local wire; it uses ARP to accomplish that. RARP maps a link layer address, such as an Ethernet address, to an IP address. As in ARP, the conversion process depends on the link layer used (Ethernet, Token Ring). The importance of RARP is diminishing, since it focuses on devices without intelligence.

3.2.5 Transport control protocol

TCP is intended for use as a reliable host-to-host protocol between hosts in a packet-switched computer communication internet. TCP is a connection-oriented, end-to-end reliable protocol designed to fit into a layered hierarchy of protocols that support multi-network communication. The TCP provides for reliable interprocess communication between pairs of processors in the host computers attached to (possibly) distinct inter-connected computer networks. TCP assumes it can obtain a simple, potentially unreliable service from the lower-level protocols. Therefore, TCP is able to operate above a wide spectrum of communication systems, ranging from hard-wired connections to packet-switched or circuit-switched networks [3].

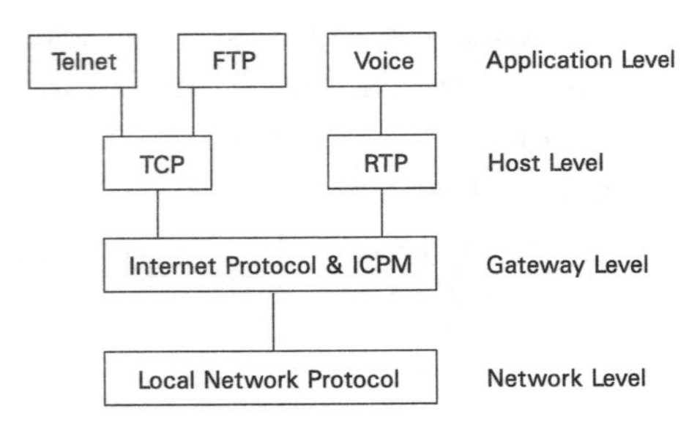

Figure 3.5 TCP protocol relationships

Figure 3.5 illustrates TCP's place in the protocol hierarchy [26]. TCP fits into a layered protocol architecture just above IP, which provides a way for the TCP to send and receive variable-length segments of information enclosed in Internet datagrams. The IP datagram provides a means for addressing source and destination TCP entities in different networks. IP also deals with any fragmentation or reassembly of the TCP segments required to achieve transport and delivery through multiple networks and gateways. IP also carries information on the precedence, security classification, and compartmentalization of the TCP segments, so this information can be communicated end-to-end across multiple subnetworks.

TCP interfaces are on one side of the protocol stack to users or application processes, and on the other side of the stack to a lower-level protocol, such as IP. This interface consists of a set of primitives, much like the primitives in an operating system provided to an application process for manipulating files. For example, there are primitives to open and close connections and to send and receive data on established connections. TCP is designed to work in a very general environment of interconnected networks. Although significant freedom is permitted for implementers to design interfaces that are appropriate to a particular operating system environment, a minimum functionality is required at the TCP/user interface. The interface between TCP and lower-level protocols is basically unspecified, except that it is assumed there is a mechanism whereby the two levels can asynchronously pass information to each other [3, 27].

TCP sends data over the transport layer, from one TCP/IP application to another (more exactly, between TCP peers). In the same way that IP uses source and destination IP addresses, TCP uses source and destination port numbers. Ports, in Internet terminology, refer to a specific protocol. Internet applications and functions are identified in TCP by a specific port number. For example, FTP is identified by port 21, Telnet is port 23, and SMTP is identified by port 25. These applications will have the same port number in any host computer on the Internet [4-21]. See figure 3.6.

TCP ensures that delivery occurs and the destination application receives the data in the correct sequence. In addition, TCP optimizes network bandwidth by dynamically controlling the flow of data between connections. If the data buffer on the receiving end starts to overflow, TCP will inform the sending end to throttle back the transmission rate.

TCP starts a timer each time it sends a message. When the destination end of a TCP connection receives a message, it sends an acknowledgment back to the sender. If

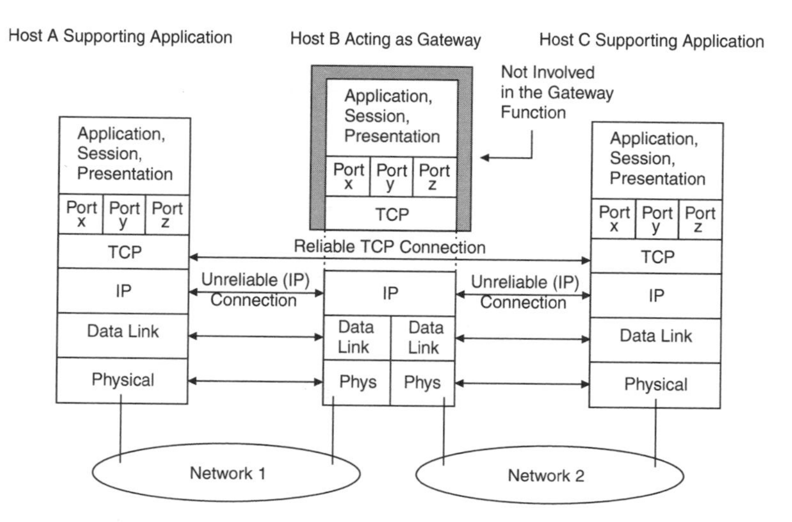

Figure 3.6 TCP environment

the timer expires before the PDU is acknowledged, then TCP will automatically resend the PDU. If TCP were to wait for an acknowledgment for each element of data it sends, then it would become very inefficient; therefore, TCP uses a sliding window. The sliding window allows TCP to send several PDUs before it pauses for an acknowledgment. In effect, TCP places an imaginary sliding window on top of the data stream and sends all the data in the window. As TCP receives acknowledgments, TCP slides the window across the data stream and transmits the next set of data.

Also, as noted previously, TCP negotiates the data flow during transmission to optimize the network bandwidth. TCP expands or shrinks the size of the sliding window according to the behavior of the network. If the network is lightly utilized, then TCP will expand the size of the send window so it can push more data into the channel at a faster rate. Inversely, TCP automatically decreases the size of the window when network congestion is high [4-21].

Each TCP segment (PDU) contains a TCP header, TCP options, and the data the segment transports. See figure 3.7. TCP has a method of identifying transmitted data and synchronizing data between sender and receiver. TCP also has a way to determine if data received are corrupted, along with a mechanism to inform the sender that these data were corrupted. TCP connections are full-duplex in that data flow in both directions at the same time. Because of this, both sides of the TCP connection must maintain two sets of sequence numbers. When a TCP sending entity establishes a virtual connection with a receiver entity, the two entities establish a system to exchange acknowledgments.

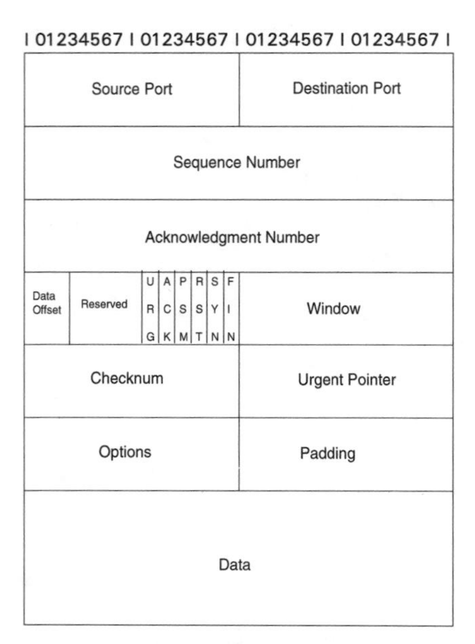

Figure 3.7 TCP segment

The TCP sender transmits a request for a TCP connection to the receiving host's transport layer entity. The TCP entity then forwards a TCP message with a synchronization flag (SYN) to the remote port to which it wishes to connect. The flag informs the receiving TCP entity that a client (sender) program wants to establish a TCP connection. The TCP message also contains a random 32-bit sequence number that the sending TCP entity uses to identify data. In turn, the receiving entity forwards a TCP segment with an acknowledgment flag and an acknowledgment number. For example, assume that the client (sender)-side TCP entity that requested the TCP connection sent a sequence number of 2,000. In response, the server (receiver)-side TCP entity stores the number 2,001 in the acknowledgment field of its initial reply message—that is, the server (receiver)-side TCP entity expects to receive data element 2,001 in the next transmission.

To establish a connection, the client (sender) entity must acknowledge the server's (receiver's) request for synchronization. This reply message from the client (sender) also sets the acknowledgment flag. In the acknowledgment number field, the client (sender)-side TCP entity stores the server (receiver)-side TCP entity's initial sequence number plus one (similar to the initial server [receiver] to client [sender] acknowledgment).

Therefore, a three-way handshake must occur to establish a TCP connection [4-21].

Step 1: Client (sender)-side TCP entity requests connection by forwarding sync request and an initial sequence number.

Step 2: Server (receiver)-side TCP entity acknowledges request for connection and requests that the client (sender)-side synchronize with the initial sequence number from server (receiver)-side.

Step 3: Client (sender)-side acknowledges server (receiver)-side request for synchronization.

In summary, when two TCP entities send and receive data, they use a sequence number to keep track of what data have been sent and received. The sequence number gives the sender and receiver a reference. To begin data transfer, the sender might send 500 bytes of information with starting sequence number 2,001. The receiver would then respond by sending an acknowledgment and the next sequence number it expects (2,502). The sequence number acts as a byte counter, tracking the number of bytes sent and received [4-21].

Applications shut down TCP connections using a two-way handshake. To close the connection, one side forwards a message with the finished (FIN) flag set, representing that it has finished sending data. The other end will then acknowledge that the connection will be closed: It will also cease to send data and will send a finished flag back as an acknowledgment of the closure; the TCP connection thus terminates [4-21].

3.2.6 TCP operation

This section expands on the operation of TCP. To provide a reliable connection service between pairs of processes over a less-reliable Internet communication system, TCP utilizes the following capabilities:

- Basic data transfer
- Reliability
- Flow control
- Multiplexing
- Connections
- Precedence and security.

Processes transmit data by calling on the module representing the TCP entity and passing data as arguments. The TCP module associated with the TCP entity packages the data in these buffers into segments and calls on the Internet module to transmit each segment to the destination TCP module. The receiving TCP module places the data from a segment into the receiving user's buffer and notifies the receiving user's upper layers (e.g., FTP). The TCP's PDU includes Protocol Control Information (PCI), which is used to ensure reliable ordered data transmission. The model of Internet communication requires that there be an Internet protocol module (layer 3 entity) associated with each TCP entity that provides an interface to the local network. This Internet module packages TCP segments inside Internet IP datagrams and sends these to a destination Internet module or intermediate gateway (router). To transmit the datagram through the local network, it is embedded in a local network packet. The Internet module may perform further packaging, fragmentation, or other operations to achieve the delivery of the local packet to the destination Internet module [3].

At a gateway (typically, a router) between networks, the Internet datagram is *unwrapped* from its local packet and examined to determine through which network the

IP datagram should travel next. The IP datagram is then *wrapped* in a packet suitable to the next network and routed to the next gateway (router) or to the final destination. A gateway (router) is allowed to break up an Internet datagram into smaller datagram fragments, if this is necessary for transmission through the next network. To do this, the gateway produces a set of Internet datagrams, each carrying a fragment. Fragments may be further broken down into smaller fragments at subsequent gateways. The datagram fragment format is designed so that the destination Internet module can reassemble fragments into Internet datagrams, regardless of how and by whom the fragments were created. The destination datagram module unwraps the segment from the datagram and passes it to the destination TCP module.

One important aspect is the Type of Service. The ToS field provides information to the gateway (or Internet module) to guide it in selecting the service parameters and communication facilities to be used in traversing the next network. Included in the ToS field is the precedence of the datagram. Datagrams may also carry security information to permit host and gateways that operate in multilevel secure environments to properly segregate datagrams for security considerations .

The host device environment TCP is supported by a software module that acts as a task, making use of the operating system. Users access the TCP much as they would access the file system. The TCP may call on other operating system functions, for example, to manage data structures. The actual interface to the network is assumed to be controlled by a device driver module. The TCP module does not call on the network device driver directly; it calls on the Internet datagram protocol module, which may in turn call on the device driver [3].

Interfaces The TCP upper-layer interface provides for calls made by the upper layers (user or user application) on the TCP entity to open or close a connection, to send or receive date, or to obtain status about a connection. The TCP Internet interface provides calls to send and receive datagrams addressed to TCP modules in hosts in the Internet system. These calls have parameters for passing the address, type of service, precedence, security, and other control information.

Reliable communication A stream of data sent on a TCP connection is delivered reliably and in sequence at the destination.* Transmission is rendered reliable through the use of sequence numbers and acknowledgments. Conceptually, each octet of data is assigned a sequence number. The sequence number of the first octet of data in a segment, called the segment sequence number, is transmitted with that segment. As dis-

---

* An acknowledgment by the remote TCP module does not guarantee that the data have been delivered to the end user, but only that the receiving TCP module has taken responsibility to do so.

cussed previously, segments also carry an acknowledgment number, which is the sequence number of the next expected data octet of transmissions in the opposite direction. When the TCP transmits a segment containing data, it puts a copy on a retransmission queue and starts a timer; when the acknowledgment for that data is received, the segment is deleted from the retransmission queue. If the acknowledgment is not received before the timer expires, the segment is retransmitted.

To manage the flow of data between TCP entities, a flow-control mechanism is employed. The receiving TCP reports a *window* to the sending TCP. This window specifies the number of octets, starting with the acknowledgment number, that the receiving TCP is currently prepared to receive [3].

Connection establishment and clearing To identify the multiplexed data streams that a TCP module may handle, TCP provides a port identifier. Since port identifiers are selected independently by each TCP module, they might not be unique. To obtain a unique address, one concatenates an Internet (IP) address identifying the TCP with a port identifier, to creating what is called a *socket,* which is unique throughout all connected networks. A connection is fully specified by the pair of sockets. A local socket may participate in many connections to different foreign sockets. A connection can be used to carry data in both directions; that is, it is full-duplex.

TCP modules are free to associate ports with processes. However, there are well-known sockets that the TCP associates only with the *appropriate* processes. Processes may *own* ports, and they can initiate connections only on the ports they own (mechanisms for implementing ownership are a local issue). A connection is specified in the open call by the local port and foreign socket arguments. In return, the TCP module supplies a local connection name by which the user refers to the connection in subsequent calls. There are several information items that must be retained about a connection. To store this information there is a data structure called a Transmission Control Block (TCB). An implementation strategy would have the local connection name be a pointer to the TCB for this connection [3]. The open call also specifies whether the connection establishment is to be actively pursued or passively waited for.

A passive open request implies that the process is willing to accept incoming connection requests from any remote TCP entity (in an active open request the TCP module is attempting to initiate a connection). In this case, a foreign socket of all zeros is used to denote an unspecified socket (unspecified foreign sockets are allowed only on passive opens), for example, service processes that provide services for unknown other processes issue a passive open request with an unspecified foreign socket. Thereafter, a connection could be made with any process that requested a connection to this local socket.

Well-known sockets are a mechanism for associating a socket address with a standard service. The Telnet process, for example, is permanently assigned to a specific socket, and other sockets are reserved for file transfer, remote job entry, and other management processes (note that the concept of a well-known socket is part of the TCP specification, but the assignment of sockets to services is outside this specification itself).

Processes can issue passive opens and then wait for matching active opens from other processes. The processes are informed by the TCP when connections have been established. Two processes that issue active opens to each other, at the same time, will be correctly connected. This capability is important for the support of distributed client/server computing, where components may operate asynchronously.

There are two cases for matching the sockets in the local passive opens and foreign active opens. In the first case, the local passive opens have fully specified the foreign socket; here, the match must be exact. In the second case, the local passive opens leave the foreign socket unspecified; here, any foreign socket is acceptable, as long as the local sockets match [3].

If there are several pending passive opens with the same local socket (noted in the TCBs), a foreign active open is matched to a TCB with the specific foreign socket in the foreign active open before selecting a TCB with an unspecified foreign socket.

As noted earlier, the procedures to establish connections utilize the synchronize (SYN) control flag and involve an exchange of messages in a three-way handshake. A TCP connection is initiated by the binding of an arriving segment containing a SYN and a waiting TCB entry, each created by an upper-layer open command. The matching of local and foreign sockets determines when a connection has been initiated. The connection becomes "established" when sequence numbers have been synchronized in both directions.

Data transmission The data that flow on a connection may be envisioned as a stream of octets. The "send" call is used. Furthermore, the sending application indicates in each send call whether the data in that call (and any preceding calls) should be immediately pushed through to the receiving application by setting the push flag. A sending TCP module collects data from the sending application and sends these data in segments at its own convenience, until and unless the push function is signaled, at which time it must send all unsent data. When a receiving TCP module detects the push flag, it must not wait for more data from the sending TCP module before passing the data to the receiving process.

The data in any particular segment may be the result of a single send call or of multiple send calls. The purpose of the push function and the push flag is to expedite data from the sending user to the receiving user. There is a relationship between the push function and the use of buffers of data that cross the TCP user interface. Each time a push flag is associated with data placed into the receiving user's buffer, the buffer is returned to the user for processing, even if the buffer is not filled. If data arrive that fill the user's buffer before a push is seen, these data are passed to the user in buffer-size units.

TCP also provides a means to communicate to the receiver of data that, at some point further along in the data stream than the receiver is currently reading, there are urgent data. TCP does not define what the user specifically does upon being notified of pending urgent data, but the expectation is that the receiving process will take action to process the urgent data quickly [3].

Precedence and security The TCP makes use of the IP type of service field and security option to provide precedence and security on a per-connection basis to TCP users. TCP modules, which operate in a multilevel secure environment, must properly mark outgoing segments with the security, compartment, and precedence [3].

3.2.7 User Datagram Protocol

UDP is utilized for a datagram-mode communication in Internets, at the transport layer level. This protocol assumes that IP is used as the underlying protocol. This protocol provides a procedure for application programs to send messages to other programs with a minimum of protocol complexity. The protocol is transaction-oriented, and delivery and duplicate protection are not guaranteed. Applications such as Simple Network Management Protocol (SNMP) use UDP.

Fields Figure 3.8 depicts the UDP PDU. Source port is an optional field (defaulted to zero). When used, it indicates the port of the sending process and may be assumed to be the port to which a reply should be addressed in the absence of any other information. The destination port has a meaning within the context of a particular Internet destination address. "Length" is the length in octets of this datagram including this header and the data (the minimum value of the length is eight). Checksum is used to provide protection against misrouted datagrams. This checksum procedure is the same as is used in TCP.

User interface A typical interface allows for receive operations on the receive ports that return the data and an indication of source port and source address; an operation that allows a datagram to be sent, specifying the data, source, and destination ports and

0–7	8–15	16–23	24–31
Source Address			
Destination Address			
Zero	Protocol	UDP Length	

Figure 3.8 UDP PDU

addresses to be sent; and the creation of new receive ports. The UDP module must be able to determine the source and destination Internet addresses and the protocol field from the Internet header.

3.2.8 Point-to-Point Protocol

The Point-to-Point Protocol (PPP—RFC 1661, 1662, and 1663) provides a method for transporting multiprotocol datagrams over point-to-point (dial-up) links. PPP is comprised of three components [3]:

1. A method for encapsulating multiprotocol datagrams
2. A Link Control Protocol (LCP) for establishing, configuring, and testing the data link connection
3. A group of Network Control Protocols (NCPs) for establishing and configuring network layer protocols

PPP is designed for simple links that transport packets between two peers. These links are assumed to provide full-duplex, simultaneous bidirectional operation, delivering packets in order. It is intended that PPP provide a common solution for easy connection of a wide variety of hosts, bridges, and routers.

Encapsulation The PPP encapsulation provides for multiplexing of different network layer protocols simultaneously over the same link. To support high-speed implementations, the default encapsulation uses only simple fields, only one of which needs to be examined for demultiplexing. The encapsulation requires framing to indicate the beginning and end of the encapsulation; see Table 3.2. The fields are transmitted from left to right. PPP's encapsulation method allows for network software to use single serial link for multiple protocols, because it uses HDLC, which is defined in the ISO standard. HDLC marks the beginning and end of each block with a special character and uses a Cyclic Redundancy Check (CRC) to detect errors in the frame.

Table 3.2 PPP PDU

Flag	Address	Control	Protocol	Data	FCS	Flag
01111110	11111111	00000011	0xXXXX (2 bytes)	0-MTU bytes	16 or 32 bits	01111110

Link-Control Protocol In order to be portable to a wide variety of environments, PPP provides an LCP capability. The LCP is used to automatically converge upon the encapsulation format options, handle varying limits on sizes of PDUs, detect a looped-back link and other common misconfiguration errors, and terminate the link. LCP negotiates optimal configuration options with each network connection it links to; thus, no matter how an administrator configures the host, LCP dynamically negotiates the optimum configuration for data encapsulation, field elimination, and frame sizes in order to optimize the throughput of transmission. Other optional facilities provided are: authentication of the identity of its peer on the link, and determining when a link is functioning properly and when it is failing [3].

Network-Control Protocols NCPs manage the specific needs required by their respective network layer protocols.

Configuration By design, the standard defaults handle all common configurations. The implementer can specify particularizations to the default configuration; these are automatically communicated to the peer without operator intervention.

Protocol field The protocol field identifies the datagram encapsulated in the information field of the packet. The field is transmitted and received most significant octet first. The structure of this field is consistent with the ISO 3309 extension mechanism for address fields. All protocols must be odd; the least significant bit of the least significant octet must equal "1" [3].

Protocol field values in the 0xxx to 3xxx range identify the network layer protocol of specific packets, and values in the 8xxx to "bxxx" range identify packets belonging to the associated NCPs, if any. Protocol field values in the 4xxx to 7xxx range are used for protocols with low-volume traffic that have no associated NCPs. Protocol field values in the cxxx to fxxx range identify packets as link layer control protocols (such as LCP).

PPP link operation overview In order to establish communication over a point-to-point link, each end of the PPP link must first send LCP PDUs to configure and test the data link. After the link has been established, the peer may be authenticated. Then, PPP must send NCP PDUs to select and configure one or more network layer protocols. Once each of the chosen network layer protocols has been configured, datagrams from each network layer protocol can be sent over the link. The link will remain configured for communication until explicit LCP or NCP PDUs close the link down, or until some external event occurs (an inactivity timer expires or network administrator intervention).

Phase diagram In the process of configuring, maintaining, and terminating the point-to-point link, the PPP link goes through several distinct phases, which are illustrated in the state diagram shown in figure 3.9. Not all transitions are specified in this diagram.

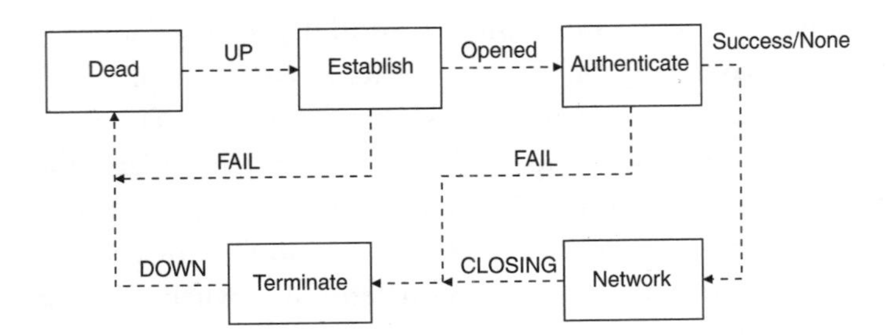

Figure 3.9 PPP phases

Link dead (physical layer not ready) The link necessarily begins and ends with this phase. When an external event indicates that the physical layer is ready to be used, PPP proceeds to the link establishment phase. During this phase, the LCP automaton is in the initial or starting states. The transition to the link establishment phase will signal an up event to the LCP automaton [3].

Link establishment phase The LCP is employed to establish the connection through an exchange of configure PDUs. This exchange is complete and the LCP opened state entered, once a configure ack PDU has been both sent and received. Only configuration options, which are independent of particular network layer protocols, are configured by LCP. Configuration of individual network layer protocols is handled by separate NCPs during the network layer protocol phase. Any non-LCP PDUs received during this phase must be discarded. The receipt of the LCP configure-request causes a return to the link establishment phase from the network control protocol phase or authentication phase. On some links it may be desirable to require a peer to authenticate itself before allowing network layer protocol PDUs to be exchanged .

Network layer protocol phase Once PPP has completed the previous phases, each network layer protocol (such as IP, IPX, or AppleTalk) must be separately configured by the appropriate NCP. Each NCP may be opened and closed at any time. After an NCP has reached the opened state, PPP will carry the corresponding network layer protocol PDUs. Any supported network layer protocol PDUs received when the corresponding NCP is not in the opened state must be discarded [3].

Link termination phase PPP can terminate the link at any time. LCP is used to close the link through an exchange of terminate PDUs. When the link is closing, PPP informs the network layer protocols so that they may take appropriate action. After the exchange of terminate PDUs, the implementation should signal the physical layer to disconnect in order to enforce the termination of the link. The sender of the terminate-request should disconnect after receiving a terminate-ack, or after the restart counter expires. The receiver of a terminate-request should wait for the peer to disconnect, and must not disconnect until at least one restart time has passed after sending a terminate-ack. PPP then proceeds to the link dead phase. Any non-LCP PDUs received during this phase are discarded [3].

3.2.9 Serial Line Interface Protocol

The Serial Line Interface Protocol (SLIP - RFC 1055) is a simpler mechanism than PPP for sending TCP/IP PDUs over direct connections and modems. The bulk of changes made were at the physical and data link layers of TCP/IP to support this function. A serial port and modem take the place of the network interface card. With SLIP, one computer can connect to another computer or to a network; or two networks can connect to each other via a router or terminal server (see figure 3.10) [3, 36]

Compressed SLIP (CSLIP) is an enhanced version of SLIP that reduces the amount of data that needs to be transmitted over the link. The result is more efficient communication, shorter sessions, and lower telephone costs. CSLIP minimizes the redundancy by sending only information that has changed in the packet. Some TCP/IP include only

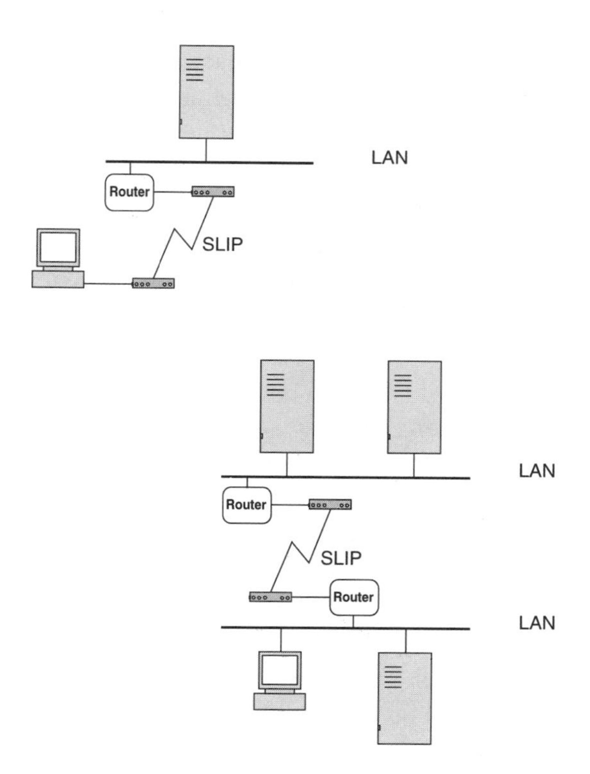

Figure 3.10 Use of SLIP

SLIP; others offer both SLIP and CSLIP. One cannot mix SLIP and CSLIP, however; both ends of the connection must use the same thing.

SLIP has its origins in the early 1980s. It is a packet framing protocol: SLIP defines a sequence of characters that frame IP packets on a serial line. It provides no addressing, packet type identification, or error detection/correction. Because the protocol does so little, it is usually easy to implement. In 1984, SLIP was implemented on 4.2 Berkeley UNIX and Sun Microsystems workstations and released to the world. It quickly caught on as an easy way to connect TCP/IP hosts and routers with serial lines. SLIP is commonly used on dedicated serial links and sometimes for dial-up purposes and is usually used with line speeds up to 28.8 Kbps. It is useful for allowing mixes of hosts and routers to communicate with one another. There are host-to-host, host-router, and router-to-router SLIP network configurations.

SLIP is one of the most popular methods used by PC and business computer users on the Internet. SLIP software is, in essence, a TCP manager, which acts as a network device driver [4-21]. SLIP can be employed to run client programs such as FTP or Telnet. In order to run these programs, they must be located on your system. With a SLIP connection, a user's computer becomes a host computer with its own IP address. Thus, a user's hardware costs for a host computer on the Internet are minimal.

SLIP is a packet framing protocol; it defines how the computer encapsulates an IP datagram before transmission over a serial line. SLIP uses specific characters to delineate the end of a SLIP frame. It uses the "End" character (ASCII 192) to delineate the end of a SLIP frame. It uses the "Esc" character (ASCII 219) to delineate that an End character or an Esc character is contained within the data inside the frame. If a user's data contain the End character, then SLIP will prefix this with an Esc, so that the receiving SLIP knows that it is not the end of the frame. If an Esc character is sent within the user's data, then SLIP will send a two-byte escape sequence. This signifies to the receiving SLIP that an Esc character is contained within the user's data inside the sent SLIP frame. Some SLIP implementations also send an End character at the beginning of the data frame to flush out any extraneous data in the frame [4-21].

SLIP relies on the TCP/IP network to provide any data error checking or correction instead of doing this itself. Both the TCP and IP protocols perform some type of error checking to detect errors.

SLIP has some deficiencies. The first is in packet addressing. SLIP typically gets its IP address from the service provider's available pool of IP addresses. In this setup, the IP address of the SLIP connection can change each time SLIP establishes a connection with the service provider. This can cause other Internet hosts to have difficulty accessing a server using SLIP. SLIP also does not include any fields that identify the PDU's destination protocol (other than IP). Thus, SLIP cannot be used in an environment where other protocols are being shared over the same cabling (i.e., you cannot use SLIP and DECnet packets over the same serial line) [4-21].

PPP resolves all the deficiencies of SLIP; thus, PPP encapsulates data with a method that allows networking of several protocols over a single serial link. As noted previously, PPP uses the LCP that network software can use to establish, configure, and test the data link connection. Both ends of the PPP connection use LCP to negotiate connection options. PPP also allows the use of different network layer protocols through its use of NCPs. PPP is also an official, adopted standard for serial link connections, whereas SLIP is not. [4-21]. Although SLIP is currently used more frequently than PPP, the eventual use of PPP should surpass the use of SLIP due to the significant advantages of PPP.

3.3 References

1 D. Minoli, *Web, Internet, Intranets*, New York, McGraw-Hill, 1997.

2 J. Davidson, *An Introduction to TCP/IP*, New York, Springer-Verlag, 1989.

3 P. DePrima, "TCP/IP, PPP, and Other Communication Access Technology for the Internet/Intranets" (*Class project, TM601, Telecommunication Technology*, Stevens Institute of Technology, January 1996).

4 P. Gaspich, "Internet Structure, Protocols, and Access" (*Class project, TM601, Telecommunication Technology*, Stevens Institute of Technology, January 1996).

5 V. Emery, *How to Grow Your Business on the Internet*; Vince Emery; Coriolis Group Books; 1995.

6 *The Internet Business Guide*; 2nd Edition; R. Resnick and D. Taylor; Sams.net Publishing; 1995.

7 *Managing Internet Information Services*; First Edition; Cricket Liu, Jerry Peek, Russ Jones, Bryan Buus, and Adrian Nye; O'Reilly and Associates, Inc.; 1994.

8 *TCP/IP Running a Successful Network*; K. Washburn and J.T. Evans; Addison Wesley Publishing Company; 1993.

9 *Internet Programming*; Kris Jamsa, Ph.D. and Ken Cope; Jamsa Press; 1995.

10 *http://www.cix.org/CIXInfo/about-cix.html*; About the Commercial Internet eXchange,.

11 *http://www.cac.washington.edu:1180/nic news/clippings/1992/03.1/index.html*; NIC News Clippings on the Internet.

12 *http://www.cix.org/Reports/registry.html*; R. D. Collett; White Paper: CIX Registry Services Strategy 2.0; April 25, 1995.

13 *http://www.cix.org/Cixtra/apr95.html#s5*; CIXtra; Technical Update: NSF Transition on Target.

14 *http://info.isoc.org/guest/zakon/Internet/History/How_the_Internet_Came_to_Be*; V. Cerf, as told to Bernard Aboba;- How the Internet Came to Be, November 1993.

15 *http://info.isoc.org/guest/zakon/Internet/History/Timeline_of_Network_History;*; S. Kulikowski II;- A Timeline of Network history; 1996.

16 *ftp://NIC.MERIT.EDU/nsfnet/transition.status*; Merit Transition of NSFNET; 1995.

17 *http://kufacts.cc.ukans.edu/cwis/writeups/internet/intro internet.html*; An Introduction to the Internet for KU Users.

18 *http://www.ans.net/Overview.html*; ANS: A Brief Overview; Advanced Network and Services, Inc.,- A ns: A Brief Overview; 1996.

19 *http://nic.merit.edu/*; Welcome to Merit Network, Inc.

20 *http://www.merit.edu/nsfnet/transition/.index.html*; NSFNET Transition.

21 *gopher://nic.merit.edu:70/1/.nsf-info*; National Science Foundation Network Proposals.

22 J. Postel, ed., *Transmission Control Protocol Specification*, Information Science Institute, 1981, p. 3.

23 J. Guyot, "World Wide Web: What Links to Databases?" p. 2.

24 Ibid, p. 3.

25 D. Minoli and A. Alles, *LAN, ATM, and LAN Emulation Technologies*, (Norwood, MAJ) Artech House, 1997.

26 Postel, *TCP Specification*, p. 12.

27 Ibid, p. 7.

28 Ibid, p. 10.

29 Ibid, p. 10.

30 Author, "Why Penn State Uses RS-232 for Internet," *Communications News*, January 1996, p. 14.

31 J. Romkey, ed. *A Nonstandard for Transmission of IP Datagrams over Serial Lines: SLIP*, Network Working Group, 1988, p. 1.

CHAPTER 4

Web Technology for the Internet and Intranets

The increase in corporate and consumer interest in the Internet, and the carry-over interest into the intranet, can be traced to the introduction of NGUI technology. NGUI* interfaces allow users to interact with the computer and its network via graphical icons, rather than using complicated commands. GUI technology, first applied strictly to PCs and client/server systems, has found its way, in the form of NGUIs, onto the Internet by way of the World Wide Web. The WWW allows users to browse the Internet as well as intranet resources via graphical interface coupled with point and click (hypertext) technology. Accessing information has never been easier. The text, hypertext, and graphical objects are retrieved from a computer housing special access software known as a Web server.

4.1 Background history

In 1989, Tim Berners-Lee submitted a proposal to the Electronics and Computing for Physics Division of the Consul Europeen pour la Recherche Nucleaire (European Laboratory for Particle Physics also known as CERN) located near Geneva, Switzerland [1]. The proposal called for a more efficient system of sharing and disseminating information to users around the world. The result of this proposal led to the birth of the World Wide Web and Web technology.

The concept of the Web was to create an environment that was capable of interfacing with all preexisting information, server protocols, and popular file formats. Users of the Web would be able to retrieve information in just about any standard format and display, animate, or play. The Web would remove the barriers of information transfer caused by incompatible computer platforms, communication protocols, and data formats.

4.2 Web technology

The Web is an environment made up of protocols and standards that allow the seamless access and transfer of information over the Internet. The Internet itself is the actual medium to transport data between two or more points. Web technology works on the premise of hyperlinking (figure 4.1). Hyperlinking is simply the linking of related information through pointers embedded in a document that point to some other (related) document. The other document need not exist on the same processor; it can exist on a processor in some other part of the world. Through the use of hyperlinking, the Web

* We define an NGUI and a network-based GUI with hyperlinks

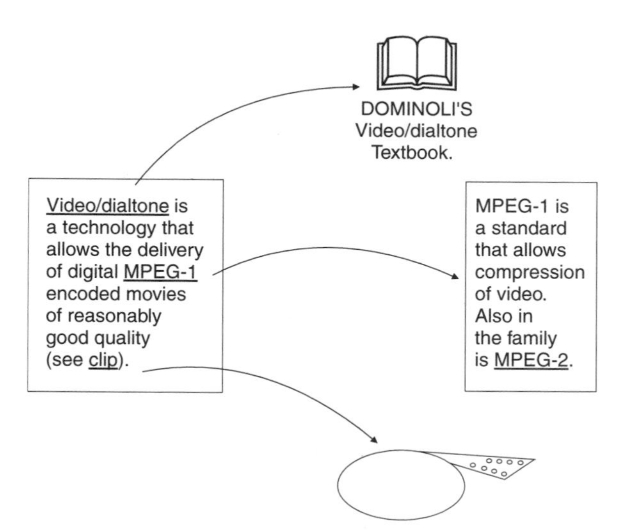

Figure 4.1 Hyperlinking

has introduced hypermedia and multimedia. Hypermedia is a combination of both hypertext and multimedia. Hypertext is text that is linked to some other related information and Multimedia encompasses the use of different types of data to represent information [2]. On the Web, one can retrieve and view information that is made up of text, graphics, and sounds. These items could be set up to act as hyperlink points to some other information. Multimedia on the Web provides the capability to download different types of information and be able to view or run them right on a computer. This information could be in the form of audio, video, text, or graphic images.

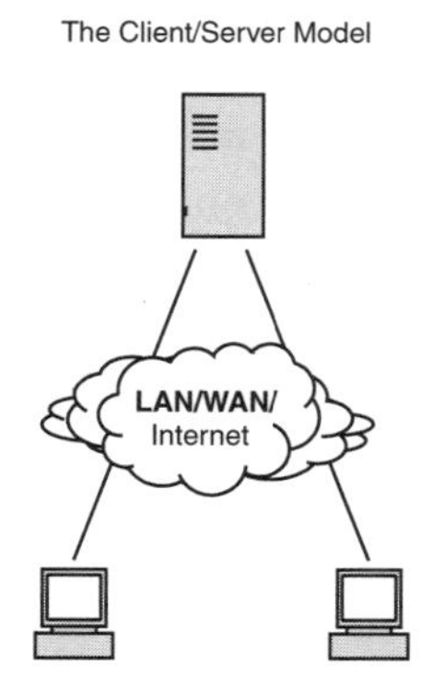

Figure 4.2 Client/server model for WWW/intranets

4.3 *The Web and the client/server model*

The Web operates on a client/ server model. This model works on the premise that Web server software runs on a processor devoted to serving documents requested by clients. The client processor receives the document, which contains specific Web formatting information, interprets this information, and displays the document on the screen for the user. Through use of the client/server model, the amount of processing performed on the server is greatly reduced, as is the ability to accommodate several clients simultaneously. See figure 4.2.

4.4 Web servers

A Web server is a system of software running on a computer that "listens" on a specific TCP port for incoming connection requests from clients. The server expects a connecting client to interface through a protocol called HyperText Transfer Protocol (HTTP). The client requesting the connection is usually a browser, such as Mosaic or Netscape, which will request some information from the server, and the server will then return the requested information to the client. Web server software is available for most of the major operating systems, such as UNIX, VM, VMS, Microsoft Windows 3.1/3.11 Workgroups, Microsoft Windows NT, Microsoft Windows 95, and Macintosh System 7. The platform on which the Web server runs is important (platform being either PC, Macintosh, or UNIX based). When choosing a platform, three basic items should be considered:

- Stability
- Performance
- A sufficient amount of Random Access Memory (RAM).

Stability is a factor, because a Web server administrator does not want the Web processor crashing due to normal processing. Up to now, UNIX has provided the most stable and flexible environment for Web servers. UNIX operating systems are designed to be full multitasking, multi/user server, which makes them perfect for Web servers. Windows NT is slowly becoming an alternative to UNIX; however, this system may need more time to prove itself as a robust Web processor. Windows and Macintosh processors are adequate for viewing Web documents. Many of today's x86/Pentium PCs and 68040 Macintosh processors are powerful enough to handle the processing and networking needs of the Web. Web servers are not generally processor intensive; they do not require high CPU speeds to operate effectively. However, they do require a sufficient amount of RAM: the more RAM provided, the more efficient the server will be in responding to client requests. At least 8 MB are required for light processing and up to 64 MB for heavy processing. Overall, high-performance Web servers (servers that are intended to cater to a multitude of users and have complex background script processes) should be dedicated processors.

4.4.1 Web server software and platforms

Today, Web servers have the greatest availability and selection for the UNIX platform. The most widely used servers for UNIX come from CERN, NCSA, and Netscape. The CERN and NCSA servers are considered public domain software, which means it is free to anyone who wants to use it. One of the problems with using public domain software is that there is little or no customer support. Both CERN and NCSA provide server-side executable scripting, image maps (clickable images), and some degree of access control. CERN offers some additional features over NCSA, such as being able to run as a proxy server and supporting document caching. CERN also provides decent documentation and is available online. The NCSA server is designed to be simple and fast. NCSA's server software is smaller than CERN's and it uses less memory. Netscape is also a supplier of Web servers. Netscape's Netsite Commerce and Netsite Communications servers are available for purchase (they are not free). They differ from CERN and NCSA in that the Netscape Commerce server provides built-in encryption security when used with the Netscape browser. Encryption becomes extremely important as companies begin to use the Internet and the World Wide Web as storefronts for the sale of goods and services. Companies must be able to provide customers some level of assurance that shopping over the Internet is safe and convenient. There are a whole host of Web servers available for just about any platform. A sample of the available servers can be found in table 4.1 [3].

Table 4.1 Web Servers

Server	Platform	Description
Alibaba	Win NT	Provided by Computer Software Manufaktur.
Apache Server	UNIX	The Apache project has produced a full-featured, general-purpose HTTP server, including nonforking server pool operation, an API for server extensions, content negotiation, extended log files, and so on. It includes the complete source code.
CERN Server	UNIX, VMS	This is the World Wide Web daemon program, full-featured, with access authorization and research tools. This daemon is also used as a basis for many other types of servers and gateways.
CL-HTTP	MAC, UNIX, Lisp	CL-HTTP is a full-featured, object-oriented HTTP server written in Common Lisp by John Mallery at the M.I.T. Artificial Intelligence Laboratory. Ports are underway to the PC.
GN	UNIX	GN is a single server providing both HTTP and Gopher access to the same data. In C, General Public License. Designed to help servers transition from Gopher to WWW.
GoServe	OS/2	This is a server for OS/2, supporting both HTTP and Gopher, from Mike Cowlishaw of IBM UK Laboratories.
GWHIS Server		This is a specialized commercial WWW server from Quadralay, Inc.
HTTP for VM		This is by R.M. Troth.
HTTPS for Win/NT	Win NT	This is a service for NT configurable using control panel.
Jungle		
MacHTTP	MAC	This is a server for the Macintosh.
NCSA Server	UNIX	This is a server for files, written in C, public domain. Many features are the same as CERN's httpd.

Table 4.1 Web Servers (continued)

Server	Platform	Description
Netsite		Netsite is a commercially supported server from Netscape Communications.
Perl Server	UNIX	This is from Marc VanHeyningen at Indiana University. Written in perl.
Phttpd Server	SunOS 5.4 (UNIX)	This is a general WWW server written in C, using features such as multithreading, memory mapping, and dynamic linking to achieve its goals of high speed, scalability, and light weight. It was written by Peter Eriksson of Signum Support AB.
Plexus	UNIX	This is Tony Sander's server originally based on Marc Van Heyningen's, but incorporating lots more stuff—including an Archie gateway.
Purveyor	Win NT, Win 95	This is Process Software's server for Windows NT and future Win32 platforms such as Windows 95.
REXX for VM		This is a server consisting of a small C program, which passes control to a server written in REXX.
SerWeb	Win 3.1	SerWeb is a WWW server that can run on the Windows 3.1 system.
Spinner		Spinner is a modularized, object-oriented World Wide Web HTTP/1.0, and HTTP/0.9-compliant server, distributed under the GPL license.
VAX/VMS server		This uses DEC/Threads for speed.
Vermeer's Personal Web Servers	Win 3.1, Win 95, Win NT	This is a 16-bit server that runs on Windows 3.1 and a 32-bit server that runs on Windows 95 and Windows NT. These are derived from NCSA 1.3; free from Vermeer.
Website	Win NT, Win 95	This is a server for NT and Win95 from O'Reilly.
Windows httpd 1.4	Win 3.1	This is a Windows server created by Robert Denny.
WN		The design of WN is based on the use of a small flat database in each directory with information about the files in that directory. Fields associated with a file include its title and may include keywords, expiration date, and any user- defined fields such as author or document ID. (User's guide, features).

4.5 Protocol predecessors of the Web

There are a variety of protocols that enable the power of the Web. However, before we discuss Web protocols it is important to understand some of the Internet protocols that influenced the Web. Internet protocols such as Archie, Veronica, X500, and WAIS (Wide Area Information Server) all contributed to the creation of the main Web protocol: HyperText Transfer Protocol, or HTTP. Two Internet protocols in particular that were instrumental in the creation of HTTP were FTP and Gopher.

4.5.1 File Transfer Protocol

File Transfer Protocol/FTP provides a standard way of accessing and transferring data from a variety of remote processors. FTP works under a dual-connection process. This means that an FTP connection involves the use of two ports. One port is used for the initial request from the client to the server, and another is used for the data transfer from the server to the client. FTP also has the characteristic of maintaining state. This means that with every request, FTP keeps a record and maintains the results of past requests. As an example of maintaining state, if a user were to set the transfer mode to binary, then any subsequent transfers would be made in binary until the transfer mode was switched back to ASCII. FTP is a powerful Internet protocol, but it has some hindering limitations when it comes to the Web. For one thing, FTP, because of its maintaining state, requires an open connection even when there are no requests. This differs significantly from the main protocol of the Web (HTTP), which only opens a connection when it has to. Another key difference is that FTP usually does not have the capability of determining what file format is to be transmitted. FTP simply transfers files, and it is the responsibility of the user to determine whether the file should be transmitted in ASCII or binary mode [4].

4.5.2 Gopher

The Internet Gopher protocol is similar to the HTTP Web protocol; however, Gopher does not support hypermedia documents. The Gopher protocol allows for searching and the linking of menu items with files. Gopher, however, only supports a small number of document types and is not able to view image graphics. Gopher is structured in a more hierarchical scheme, as opposed to the Web's interlinking of numerous objects. Using Gopher is much like using a menu system. A user would traverse the menu until the desired information is located [5].

4.6 Uniform Resource Locators

A Uniform Resource Locator (URL) sometimes referred to as a Universal Resource Locator), is simply a means of identifying the location of any particular resource on the Internet. URLs typically provide the protocol in which a request is made, as well as the physical address or location of the desired resource. The protocol used may determine the specific format of the URL string; however, a large number of protocols have adopted a common URL format. This common format is known as the Common Internet Scheme Syntax (CISS).

The CISS format is as follows: `schemename://username:pasword@host:port/path,` where schemename is the name of the protocol being used (see table 4.2). The symbol "`://`" is the primary syntax, which identifies a URL's compliance with the CISS. URLs originated as a requirement from the Uniform Resource Identifier (URI)

Table 4.2 Protocol Name/Ports

Scheme	URL Format\	Defaults
File	file://host/path	None
FTP	ftp://user-password@host:port/path	Username=anonymous Password=user's email address Port=21
HTTP	http://host:port/path?searchpath	Port=80
Gopher	gopher://host:port/path	Port=70
Mailto	mailto:local-address@host	None
News	news:newsgroup-name news:message_id	None None
NNTP	nntp://host/newsgroup-name/article-number	Port=119
Prospero	prospero://host:port/hsoname;field =value	Port= 1525
Telnet	telnet://user:password@host:path	Port=23 (If omitted, user is prompted for user name and password.)
WAIS	wais://host:port/database	Port=210

specification (RFC 1630). The goal of the URI specification was to come up with a comprehensive technique of addressing information over the Internet. This could only be accomplished by devising an open naming scheme that does not make any assumptions about the scheme being used. This type of open scheme is known as an *extensible format*. URLs may include the following characters in addition to alphanumeric characters: dollar signs (`$`), hyphens (`-`), underscores (`_`), periods (`.`), plus signs (`+`), exclamation points (`!`), asterisks (`*`), quote marks (`'`), commas (`,`), and left "`(`" and right "`)`" parentheses. Special characters, such as tabs and new lines, must be encoded with their specific escape code sequence, as follows:

Tab	`%09`
New line	`%0a`
Space	`%20`
At sign (@)	`%40`
Ampersand (&)	`%26`
Question Mark (?)	`%3f`

4.7 HyperText Transfer Protocol

The primary protocol for information transfer on the Web is HTTP. HTTP operates in the following manner:

1 The client establishes a connection to the Web server.
2 The client requests a specified document from the Web server.
3 The server responds to the client with a status code and the document (if available).
4 The client or the server disconnects.

The HTTP protocol is rather simple for several good reasons; the first reason is speed. Quick response is a primary goal of HTTP. One of the ways HTTP achieves quick response time is by being a stateless protocol. A stateless protocol does not maintain the results of past requests. Each request is treated as a new request with no knowledge of the past—unlike FTP, which remembers the results of previous requests. HTTP is only allowed one request per connection, and each time a request is made, a new connection must be established. Another reason is extensibility. By keeping the protocol simple, it makes it easy to provide support for new data formats. HTTP operates through the execution of valid methods. A method is defined as an action to be performed on a particular resource. Most Web sites use one of two versions: 0.9 and 1.0. HTTP 1.0 is a more robust implementation than 0.9. HTTP 1.0 supports transaction headers and incorporates new methods that did not exist under HTTP 0.9. There are plans to redesign HTTP into a new version named: HTTP-NG, where the NG stands for next generation. HTTP-NG will address some of the speed issues with HTTP 1.0. HTTP 1.0 supports seven methods. These methods are listed in table 4.3 [6].

Table 4.3 HTTP Verbs

Method	Description
GET	GET requests an object that is identified by the URI from a server. GET will request the contents of an object that is either a document or a file. If the object is an executable program or script, GET requests the resulting output. If the object is a database query, GET will request the results of that query. The GET method is run each time a request is made to link to another Web page or document.
HEAD	HEAD requests the server to send an object's meta-information, and not the object itself. Many browsers that use caching use the HEAD method as a way to increase response time. If a particular document is already cached on the client, the browser will use the HEAD method to find out what the last modified date of the document is. If it has not changed, the browser will pull up the cached document locally instead of having to retrieve it again.
PUT	PUT creates a new resource or updates an existing one. PUT transfers data to the server and makes it available through a specified URI. If data already exists at the URI, these data are treated as a new version; otherwise, if these data are new, the server attempts to create a new resource at the specified URI.
DELETE	DELETE issues a request to the server to remove a specified resource.

Table 4.3 HTTP Verbs (continued)

Method	Description
LINK	LINK requests the server to add link relationships to objects specified by the URI. Link relationships may have the following properties: • One resource is a continuation of another. • One resource is a predecessor of another. • One resource is the parent of another (hierarchy tree). • One resource is the creator of another.
UNLINK	UNLINK issues a request to the server to remove link relationships from the object specified by the URI.
POST	POST allows a user to transfer data from the client to the server. Once the data is received on the server, the server will process it. Data in this context are usually the contents of an HTML form.

A client may make use of one or more of these methods; however, the server only has to support the GET and HEAD methods to be HTTP 1.0 compliant.

The use of headers is important to a Web server. There are two basic types of headers: request headers and response headers. Request headers inform the server how to interpret a request. The type of request headers available are:

Accept:	Accept names a list of content types that the client will accept in a response.
Accept-Charset:	This specifies the character sets that the client is able to accept in a response.
Accept-Encoding:	Specifies a list of content-encoding types the client is able to accept in a response.
Accept-Language:	This contains a list of languages that the client is able to accept in a response.
Authorization:	The server uses this information to verify that the client has authorization to make the request.
From:	Contains the requester's email
If-Modified Since:	This is used with the GET method to determine whether an object has been modified since the supplied date. If it has, then the server will send the object; otherwise, the server will return a status code of 304.
Pragma:	The client tells the server to behave as requested. Currently, HTTP 1.0 only supports a value of "no-cache," which instructs the server to request a document despite whether a server is chaching or not.

Referrer:	This identifies the last oblect link of a linked request (i.e. when the user links to next.web.srv.com from the paper.web.srv.com page, the referrer header would contain paper.web.srv.com).
User-Agent:	This tells the server what client application is being used (i.e. Microsoft Internet Explorer, NCSA Mosaic, Netscape).

Response headers are header information sent to the client once a request has been made. The available response headers are:

Public:	The server may specify a list of nonstandard methods supported by the server.
Retry-After:	This designates the amount of time before the requested resource is available again.
Server:	This identifies the server software.
WWW-Authentication	This is used to advise the client that the requested URI is secured and that the requester is to be authenticated. The WWW Authentication response header includes the authentication scheme to be used.

HTTP 1.0 response header strings observe the following convention:

- Entity-Body
- Entity-Header
- General-Header
- HTTP-Version Status-Code Reason-Phrase
- Response-Header

The Entity-Body contains the requested object itself (if available).

4.8 Security for Web Servers

The increase in using the Web for the sale of goods and services has promoted the need for security on the Web. There are a variety of security measures available; however, one of the focal points is to provide security built into the HTTP protocol. Security protocols such as Secure HTTP (SHTTP), Shen, and Secure Sockets Layer (SSL) are attempting to supply security by implementing authentication and encryption technology.

Both Secure HTTP and Shen are attempts to provide security for HTTP transactions. These protocols are basically extensions to the HTTP 1.0 protocol, providing authentication and encryption security between the client and the server. SHTTP is an encryption algorithm proposed by CommerceNet. CommerceNet is a consortium that has a vested interest in developing and using the Internet for commercial purposes. SHTTP is a higher-level protocol that currently works only with the HTTP protocol. On the server side, SHTTP is currently being implemented on the Open Marketplace server (marketed by Open Market, Inc.). On the client side, SHTTP can be found on the Mosaic browser (Enterprise Integration Technologies) [7].

Secure Sockets Layer (SSL) is an encryption algorithm proposed by Netscape Communications. SSL is a lower-level protocol and is used to encrypt communication within higher-level protocols, such as HTTP, NNTP, and FTP. The SSL protocol has the capability to do server authentication (verifying the server to the client), data encryption, and client authentication (verifying the client to the server). The Netscape browser and some versions of the Netscape servers are the only applications that have implemented SSL at press time. SSL employs RSA cryptographic techniques to implement data encryption. RSA is a variable-length public key cryptographic algorithm, which uses a mathematical formula to encrypt data. The length of the key can vary between 40 and 1,024 bits. Netscape browsers sold in the United States (for use in the United States) support key lengths of 128 bits. Any Netscape secure server can support key lengths up to 128 bits. What is interesting here is that Netscape browsers exported from the United States must support a 40-bit key length in order to comply with U.S. export laws.

4.9 Common Gateway Interface

As the Web began to grow, it became apparent that there was a need to standardize the way Web servers and server-side executables provided one another input and output. Without such a standard, each server may support server-side executables in a different manner, which leads to problems with portability and interoperability. This standard, developed by the NCSA, is known as the Common Gateway Interface (CGI). The CGI standard provides two ways in which the server and the server-side executable can communicate (see figure 4.3):

1. The way the server passes input to the executable
2. The way the executable passes output back to the server.

In order to accommodate these two standard practices, CGI uses a defined set of environment variables to facilitate the passing of information between the server and the server-side executable (see table 4.4). Since most, if not all, scripting and programming languages can read environment variables, this has proven to be a good method of passing input and output information.

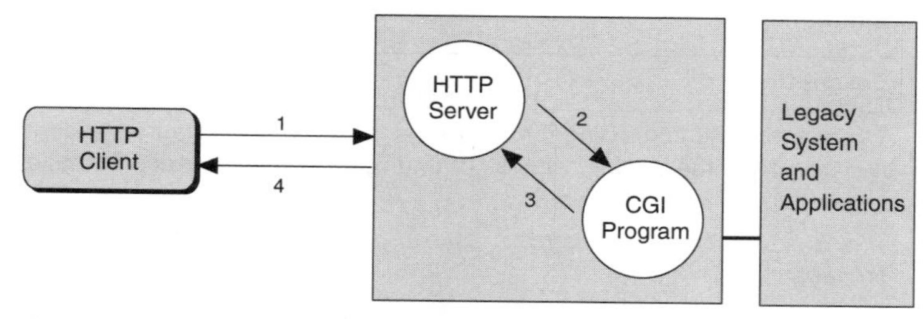

1) The client connects to the HTTP server and sends a request.
2) The server sets its respective environment variables and launches the CGI script.
3) After completing its processing, the CGI script passes the results back to the HTTP server.
4) The HTTP server sends the response to the client.

Figure 4.3 CGI process flow

Table 4.4 CGI Variables

Environment Variable	Description
AUTH_TYPE	This is a variable containing the authentication scheme used in the request. Currently there is only one value defined: *Basic*. The *Basic User Authentication Scheme* is the authentication scheme defined under HTTP/1.0.
CONTENT_LENGTH	This is a variable containing the number of bytes submitted with the request.
CONTENT_TYPE	This is a variable containing the MIME type of the data that was submitted with the request.
GATEWAY_INTERFACE	Variable which contains the CGI version supported by the server. This variable is of the form CGI/version.
HTTP_	If a header field is submitted that does not correspond to any of the standard environment variables, an HTTP_[field_name] environment variable is created.
PATH_INFO	This is a variable containing the piece of the URI that follows the CGI script to be run. In the URL `http://www.telcom.com/cgi/run_cgi/new/args`, the "*/new/args*" string would populate the PATH_INFO variable.
PATH_TRANSLATED	This is a variable containing the actual (physical) path of the information contained in the PATH_INFO variable. In the URL `http://www.telcom.com/cgi/run_cgi/class.html` the "`/class.html`" string would translate into `/web/local/bin/html/class.html`.
QUERY_STRING	This is a variable containing the search criteria submitted with the request.
REMOTE_ADDR	This is a variable containing the *IP address* of the host making the request.
REMOTE_HOST	This is a variable containing the *host name* of the host making the request. This variable is only set if the server can inverse-query the IP address of the remote host. If it cannot, this variable is left empty.

Table 4.4 CGI Variables (continued)

Environment Variable	Description
REMOTE_IDENT	This is a variable containing the *user id* as retrieved from the user's identified server. If the server cannot communicate with the user's identified server, the variable is left empty.
REMOTE_USER	This is a variable containing the *user id* provided via the authentication process.
REQUEST_METHOD	This is a variable containing a string that indicates the method used in the client's request. The REQUEST_METHOD can take the value of the following: GET, POST, HEAD, PUT, DELETE, LINK, and UNLINK.
SCRIPT_NAME	This is a variable containing the piece of the URI that identifies the CGI script. In the URL `http://www.telcom.com/cgi/run_cgi`, the `"/cgi/run_cgi"` string would populate the PATH_INFO variable.
SERVER_NAME	This is a variable containing the *host name* or *IP address* of the server.
SERVER_PROTOCOL	This is a variable containing the HTTP version supported by the server. This variable is of the form HTTP/version.
SERVER_SOFTWARE	This is a variable containing the *name* and *version* of the HTTP server software. This variable is of the form *name/version*.
SERVER_PORT	This is a variable containing the port number on which the Web server is communicating.

One may occasionally hear references to CGI scripts. The Common Gateway Interface standard does not specify any particular programming or scripting language to produce CGI-compliant programs. CGIs are usually written using a scripting language such as sh, bash, or perl. However, this does not mean that programming languages such as C, C++, Visual Basic, and Visual C++ could not be used.

An example of a CGI script written for a UNIX system in Bourne shell follows.

```
#!/bin/sh
  FINGER='which finger'
  echo Content-type: text/html
  echo
  if [ "$QUERY_STRING" = "" ]; then
    echo "<TITLE>Finger Gateway</TITLE>"
    echo "<H1>Finger Gateway</H1>"
    echo "<ISINDEX>"
    echo "This is a gateway to \"finger\". "
    echo "Type a user@host combination in your browser's search
      dialog.<P>"
  else
    echo "<PRE>"
    $FINGER "$QUERY_STRING"
    echo "</PRE>"
```

This generates a page of HTML that allows the user to enter the user name of the person to query, unless it's called with a user name in QUERY_STRING, in which case it executes the UNIX finger command using QUERY_STRING as a parameter and then returns the result to the user [9].

CGI scripts are useful for handling much of the back-end processing associated with an interactive client/server model. CGIs are extremely useful for creating search engines, filters, or gateways to other server-side protocols and interfacing with a locally resident database. Images that a user can click on are supported by CGI scripts. The part of the image the user clicks on is encoded, and the image map coordinates are sent to the server script. These coordinates are then used to determine the associated URL to return to the client. Another use for CGI scripts is generating documents *on the fly*. This is a powerful use of CGIs, because a Web document can be tailored based on input from the user. It is important to keep in mind that CGIs work on the server side; therefore, one must take into consideration the type of processing the CGI will be doing and whether or not the server processor can handle the load.

4.10 Web server access

Accessing a Web server is accomplished through a variety of text and graphical software tools known as Web browsers. There are a variety of browsers available on the market today. Text browsers such as Lynx and graphical browsers such as NCSA's Mosaic and Netscape's Navigator have opened the Web to a whole new world of users.

4.10.1 Browsers

Browsers provide a universal way of communicating with a variety of servers. World Wide Web browsers use a simple graphics-based user interface instead of entering command strings. WWW browsers allow a user to just point and click. WWW browsers provide a way for information to be organized in a non-linear manner using links. Through the use of links, any item of information (a word, phrase, or image) can function as a link to any other item of information. Underneath every link is a URL, which tells the browser where to find the resource pointed to by that link.

There is a variety of browsers available. Thus far, Netscape's Navigator is the most popular. Some of the other browsers available are NCSA's Mosaic, Lynx (text-only browser), WebExplorer, Air Mosaic, and Microsoft's Internet Explorer. Table 4.5 shows a breakdown of a percentage of hits based on a sample size of 260,000 requests registered at the `http://www.netgen.com` Web site (information was gathered on April 28, 1995).

Table 4.5 Use of Various Browsers

Browser	Percentage of Hits (%)
Netscape	66
NCSA Mosaic	10
Lynx	5
WebExplorer	3
Air Mosaic	2
Other	13

4.10.2 Access from the network

Network access to Web servers is achieved using TCP/IP. As discussed in Chapter 3, TCP/IP provides for the routing of information from one node to another, regardless of whether the node is running the same operating system (i.e., UNIX, DOS/Windows, etc.). TCP/IP can be used to interconnect personal computers forming user LANs.

4.10.3 Point-to-Point Protocol and Serial Line Interface Protocol

Access to the Internet is accomplished in several ways. Access can be attained via a company's LAN environment or by using a modem connection. Within the LAN environment, using an Ethernet connection is the most efficient, since the Internet uses an Ethernet protocol. Users who use a modem for dial-up access will either dial into a shell account, or a Point-to-Point Protocol (PPP) or Serial Line Interface Protocol (SLIP) account. As discussed in Chapter 3, SLIP and PPP are dial-up protocols that allow a user's PC to become a remote client of the network. This should give the user all the network connectivity he or she would have if directly connected to his or her LAN. Both SLIP and PPP can run at 14.4 Kbps or better (depending on the capabilities of the dialup and PC modem). Access to the World Wide Web accomplished through browsers such as Netscape and Mosaic works most effectively in dial-up mode with SLIP or PPP.

4.10.4 Firewalls

One of the more effective ways of protecting one's Web server from attacks is to implement a firewall. An Internet firewall is a security device that allows limited access to one's site from the Internet, as well as permiting approved traffic in and out of one's local site. This type of security measure allows one to select applicable services necessary to one's business and lock out any services that may be a potential security risk. This type of control allows an administrator to preselect the services covering his or her needs, while preventing unwanted visitors from compromising the systems. Firewalls not only protect internal networks from untrusted networks (either internal or external), but they may also be used to segment internal LANs, based on operational functionality. This segmentation would be useful in keeping sales personnel from gaining access to development/architecture systems. Firewalls are capable of screening virtually every one of the seven layers of the protocol architecture model.

4.11 HyperText Markup Language

This section introduces the concept, which is expanded upon in the next chapter.

Web services of today primarily use a language known as HyperText Markup Language (HTML) to define how data will be presented to a user. The HTML language is a standard defined in the Standard General Markup Language (SGML) guidelines. HTML provides a way of defining areas of text and associating those areas with a specific format tag. These tags provide a functional description of an area of text, not how an area of text should be displayed. Browsers have no restrictions on how to interpret a specific HTML tag. As an example, one may set an HTML tag to render an area of text as a first-level header. Depending on the browser a user has, this first-level header may be interpreted and displayed to the user in different ways. One browser may center first-level headers and another may underline them. This method of setting format tags is known as semantic markup. In general, HTML does not support physical markup — that is, HTML cannot be used to set text to 24-point bold. One of the powerful features of HTML is its ability to support hypertext links between documents and objects. A user could create a Web document and link into various other Web documents, which then in turn could have embedded links to other Web documents. These links of information are what form the World Wide Web.

The latest version of HTML (version 3.0) supports the basic text outputs with hyperlinking and forms. Version 3.0 also supports tables, mathematical equation support, and improved layout control. Some browsers such as Netscape have added extensions to HTML to give users more capabilities when rendering Web pages through the Netscape Navigator browser. HTML was designed to be browser independent. This implies that as long as HTML documents are coded to specification, any browser should be able to render them. Coding HTML documents to specification becomes significant, especially when dealing with different browsers. Many of today's browsers may have subtle differences in how they support HTML. Using HTML coding techniques that work for one particular browser may produce undesirable results for another. If a browser does not understand a tag, it will ignore it. Figure 4.4 depicts a simple HTML document [10].

```
<TITLE>The simplest HTML example</TITLE>
<H1>This is a level-one heading<H1>
Welcome to the world of HTML.
This is one paragraph.<P>
And this is the second.<P>
```

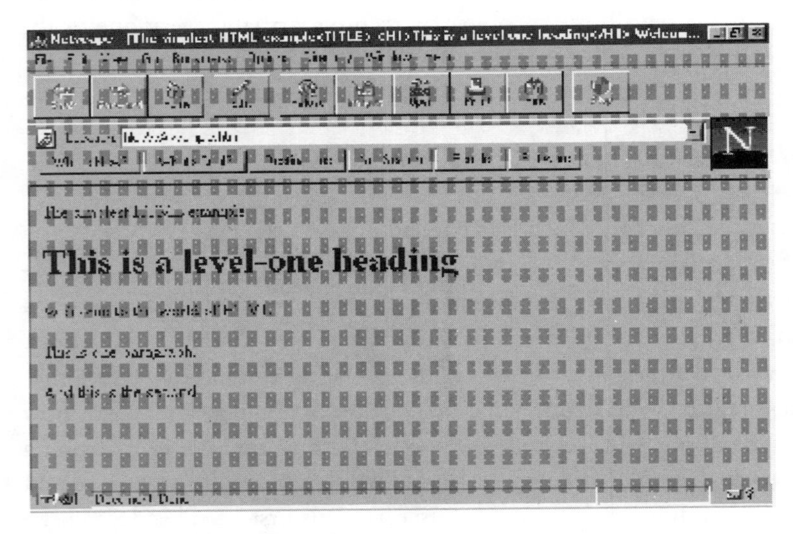

Figure 4.4 Simple HTML document

Every HTML document contains at least two elements, where an HTML element is an identifier for a particular component of an object. These elements are comprised of three parts:

1. Beginning Tag. A beginning tag denotes the element being used. Beginning tags use the following convention: “<”, then the tag name, and then a “>”.
2. Content. The content is defined as anything that appears between the beginning and ending tags.
3. Ending Tag. Ending tags use the following convention: “</”, then the tag name, and then a “>”.

HTML documents, as a general rule, should begin with an HTML element which identifies the content as HTML. The HTML element has two parts: the head and body elements. The head element contains information about the content of the body. The body contains the principal content of the document. Together, head and body are known as document structure elements (see figure 4.5).

HTML Document Structure

<HTML>

<head>

<head>

<body>

<body>

</HTML>

Figure 4.5 Document structure elements

HTML code is written and viewed in plain ASCII text. Any word processor or text editor can be used to write an HTML document (as long as it is saved as text). Some of today's word processors and text editors support tools to create HTML documents without actually having to write HTML code. Word processors such as Microsoft Word and text editors such as Emacs will autoconfigure a document with the appropriate HTML tags. These tools will usually create a basic HTML document; therefore, an author may have to edit the document to achieve a desired layout. HTML is not case sensitive; therefore, <head> is the same as </HEAD>.

4.11.1 Clickable image maps

Image maps are a powerful means of displaying information and creating interesting graphical link points. One should remember, however, that not all browsers are graphical. There is a significant number of users who use text-only browsers. Obviously, text-only browsers are not capable of rendering graphical images. Text-0nly browsers will typically display graphical images as a text label, such as `[ISMAP]` or `[IMAGE]`. Clickable image maps should be used carefully. Images take extra time and processing power to download and display. Clickable images also require some extra processing power on the server end. The server needs to run a CGI script to determine the URL associated with the image. A good practice when using clickable image maps is to supplement the image(s) with a text-based hyperlink(s) that will access the same locations (figure 4.6) [11].

[Prev] [Next] [Home] [Search] [Mailto] [Affiliates:] [Security First Network Bank] [Five Paces, Inc.]

Figure 4.6 Clickable locations

4.12 The future of Web server technology

Advanced uses for Web technology are just now being realized. This technology allows users to display objects graphically and has opened up a whole new dimension to commerce. As discussed previously, this technology is now entering the intranet environment. The efficiency of HTTP and the ease of creating Web documents with HTML, make the Web open to all users, both technical and nontechnical. The future can only hold more efficient and secure Web server protocols that will be able to support worry-free commerce and better handling of objects such as 3-D imaging and video. The next generation of Web servers will likely provide enhanced support for virtual reality.

4.13 References

1. "How the Web Was Created", *http://www.hcc.hawaii.edu/guide/www.guide.html#t1*.
2. "Entering the WWW: A Guide to Cyberspace", *http://www.hcc.hawaii.edu/guide/www.guide.html*.
3. "World Wide Web Server Software", *http://www.w3.org/hypertext/WWW/Servers.html*.
4. "File Transfer Protocol (FTP)“, *http://www.w3.org/hypertext/WWW/Protocols/rfc959/2_Overview.html*.
5. "Introduction to the Internet", *http://uu-gna.mit.edu:8001/uu-gna/text/internet/notes/gopher*.
6. "HTML Specification Version 3.0", *http://www.hp.co.uk/people/dsr/html3/CoverPage.html*.
7. "Secure HTTP", *http://www.commerce.net/information/standards/drafts/shttp.txt*.
8. "Secure Socket Layer", *http://home.netscape.com/info/SSL.html*.
9. "An Introduction to the Common Gateway Interface", *http://www.utirc.utoronto.ca/CGI/cgi1.html*.
10. "The Minimal HTML Document", *http://www.ncsa.uiuc.edu/General/Internet/WWW/HTMLPri*mer.html
11. "SecureWare, Inc.", http://www.secureware.com.

CHAPTER 5

Software Tools in Support of Intranet WWW Development

This chapter* covers the software tools that are used to develop applications residing on the servers on the intranets (and on the Internet). The very fact that these tools are powerful and allow the delivery of hypermedia, multimedia, graphics, voice, and data implies that a robust communication infrastructure will be required by corporations and organizations. As indicated in chapter 1, by the year 2000, we believe that many corporate intranets may well look like this: `HTML/HTTP/TCP/IP/MPOA`. The issues related to MPOA (and ATM) are treated in the following chapters. This chapter focuses on HTML and related systems. Although some of the tool information will be time-dependent, these products and these principles are expected to significantly influence the deployment of intranets well into the next decade.

5.1 Languages

Predominant among Web development tools are languages used to access/retrieve information. This section describes three Web languages: HTML—the current standard language, JAVA—the emerging de facto language, and VRML—an up-and-coming language dealing with three-dimensional graphics.

5.1.1 HTML

History and Foundation HTML is today the most widely used language for WWW homepage development. HTML is an application of the Standardized Generalized Markup Language (SGML), an ISO standard (No. 8879) written in 1986. HTML has been used on the Web since 1990 and has a very active revision history. Even though several browser vendors have added their own unique extensions to HTML, the World Wide Web Consortium (W3C) officially maintains HTML specifications. Once HTML specifications are approved by the W3C, they go through the Internet Engineering Task Force (IETF) and become an official RFC for the Internet. To date, HTML has the following W3C version history:

- HTML (anything prior to version 2.0 was an informal version)
- HTML2.0
- HTML+, (an extension of HTML2.0)
- HTML3.0.

* D. Kinsky [1] contributed to this chapter.

HTML's growth can be explicitly attributed to the growth of the Web. It is a very simple yet scalable language which can be used for information exchange on virtually any platform including the following:

- GUI environments such as Windows, Macs, X Windows
- Text-only environments such as VT-100
- Text-to-speech devices.

HTML is predominantly a text markup language. It supports the ability to hyperlink to other Internet resources based on clicking on a word/phrase on the screen. It is a revolutionary change to the prior *Telnet* and *FTP* protocols that defined previous Internet access. Some recent versions of HTML even support object hyperlinking—for example, a user can view a bitmap detailed picture, put the mouse on the portion of detail of interest—and HTML will pull up the pertinent piece of information.

Language specifics HTML documents are in plain text (i.e. ASCII) format. HTML uses markup tags to tell the Web browser how to display the text (or image) within the tag. The syntax of markup tags is as follows.

```
<Tag1> Some text or other information </Tag1>
<Tag2> Some text or other information </Tag2>
```

Note the "/" preceding the close of the tag. This tells the browser that the end of the tag has been set.

Table 5.1 lists some basic HTML markup tags.

Table 5.1 HTML Markup Tags

Tag	Description
Title	The `<title>` tag represents the title of the whole document. This will often be used in WAIS (index) searching. Text following the `<title>` is also deployed at the top of the window.
Headings	The `<Hy>` tag represents six levels of headings (numbered `H1` through `H6`). Headings are typically bolder and/or larger than normal text. H1 is the most prominent; `H6` is the least prominent.
Paragraphs	The `<P>` tag represents the beginning of a new paragraph. This is the equivalent of a carriage return and line feed and another line feed (e.g., a space between the old text and the new paragraph).

Table 5.1 HTML Markup Tags (continued)

Tag	Description
Anchor	The <A> tag represents the anchor tag and is by far the most significant feature of HTML. It allows links to other regions of text within the same document or in another document. The <A (note that right angle bracket is missing) is often followed by the argument "HREF"– which points to another document—for example the HTML code: `<A HREF="http://home.netscape.com">` Netscape's home page</A> <u>produces Netscape's home page,</u> and when the user clicks on the underlined phrase it opens an `http://` connection to Netscape's home page.
Unnumbered Lists	The <UL> tag represents an unnumbered list. Before the </UL> closing tag, several items can be listed, provided that they are preceded by the subtag <LI> (note that no </LI> is needed)—for example, the HTML code <UL> <LI> apples <LI> bananas </UL> produces apples bananas The <LI> items can contain multiple paragraphs separated by <P> tags
Numbered Lists	The <OL> tag represents numbered lists or ordered lists. Before the </OL> closing tag, several items can be listed, provided that they are preceded by the subtag <LI> (note, no </LI> is needed)—for example, the HTML code <OL> <LI> apples <LI> bananas </OL> produces 1. apples 2. bananas The <LI> items can contain multiple paragraphs separated by <P> tags.
Definition Lists	The <DL> tag represents the definition list. Before the </DL> closing tag, several definition terms can be listed, provided that they are preceded by the subtag <DT>. Within each definition term there can exist a definition definition which is tagged as <DD> (note that no </DT> or </DD> closing tag is needed)—for example, HTML code: <DL> <DT>ORG CODE <DT>LOC CODE <DD> ORG CODE The Organization Code identifies an employee's organization code in the hierarchy of the company. LOC CODE The Location Code identifies an employee's location code in the geographical hierarchy of the company. Note, the <DD> tags can contain multiple paragraphs separated by <P> tags.

Table 5.1 HTML Markup Tags (continued)

Tag	Description
Nested Lists	Nested lists are an adaptation of the unnumbered list tag—for example, the following: HTML code <UL> <LI> A few European countries: <UL> <LI>France <LI>Germany <LI>Italy </UL> <LI> One North American country: <UL> <LI> Canada </UL> </UL> produces A few European countries: –France –Germany –Italy One North American country: –Canada
Preformatted Text	The <Pre> tag represents text that should appear exactly as it is formatted in the file (for fixed-width font, spaces, and line breaks). For example, the HTML code: <PRE> cd $HOME/bin echo "hello" |a.out >out.txt cat out.txt </PRE> displays exactly as cd $HOME/bin echo "hello" |a.out >out.txt cat out.txt Note that in a regular stream of text, the browser ignores the line breaks until it reaches the end of the line. Here, a line break is preformatted.
Inline Images	The <IMG> tag represents an inline image that can be displayed (e.g., a .gif or .bmp file). The <IMG is followed by the "SRC=" command, which indicates the location of the image. For example: <IMG SRC="UpArrow.gif" ALT = "UP"> tries to display the UpArrow.gif image (location is current directory), and if the browser cannot do this (e.g., it's a text-only environment such as a VT100 session), then it displays the text "UP" instead.
Head and Body	The <HEAD> ...... </HEAD> and <BODY> ..... </BODY> tags separate the document into introductory information and main text. They do not affect the appearance of the formatted document–rather, they are tags used in some browsers to download just the HEAD portion first and then let the user decide whether or not to download the BODY portion.
Some Text Formatting Tags	Additional formatting tags are as follows. Their definitions are given within the tag. <I>...*italics*....</I> <B>...**bold**.....</B>

Revision history

HTML 2.0 specification HTML 2.0 is currently a baseline specification according to the World Wide Web Consortium. It is also an official Internet RFC (1866) as of September 1995. The major additions to the basic HTML markup tag listed in table 5.1 are presented in table 5.2.

HTML+ and HTML3.0 specification HTML+ and HTML3.0 are not true specifications. A 3.0 specification was written in March 1995, but it expired in November 1995. There are several references to items that will be part of HTML3.0, according to W3C. HTML3.0 is currently an unofficial specification. Table 5.3 lists some of the enhancements of HTML 3.0 over HTML2.0.

Table 5.2 HTML Markup Tags

Tags	Description
Added HyperLink Features	In <IMG> elements, the *x,y* coordinates follow the mouse on the screen. The HTML author can capture the *x,y* coordinates and hyperlink to another area based on the *x,y* coordinates.
Forms Support	Initial forms support allows for a <form> tag with input types text password, checkbox, radio button, image, pixel (where the mouse returns an x,y coordinate map of the image), submit, and reset.
Handling Newline Characters\	HTML2.0 is required to recognize newline characters as: • Carriage Return, • Line Feed • A combination of CR and LF
Security	When the user submits a form (or any other data sent on to the network), a security message stating possible exposure pops up on the screen and prompts the user to clarify the submission request.

Table 5.3 HTML 3.0 Enhancements

Tags	Description
Inserting/Embedding Multimedia Objects	True object plug-in capability instead of just HTML2.0 static images is provided. One current tag is <Applet> for Java applets. Other tags might include <EMBED>, <FIG>, and <INSERT>.
Style Sheets	HTML authors can define specific styles and use those styles within their Web page(s). Style types can be associated as follows: • User-applied style sheets (the author lets the user select the preferred style) • Implicit associations (the style sheet is implied by a URL address and retrieved as needed) • Explicit associations (the author explicitly defines one or more style sheets to a Web page)

Future of HTML There is an effort underway in the W3C to enhance HTML in a number of ways. Table 5.4 presents some of the possible future implementations.

Table 5.4 Possible Future HTML Enhancements

Extended Alphabet Set	Currently, HTML is limited to 192 characters, based on the Latin Alphabet Number 1 (ISO 8859-1). While that is sufficient for western European languages, it is insufficient for an international audience. There is currently an Internet draft dated 11/22/95 that seeks to expand the number of characters in HTML. Expiration of the draft is scheduled for May 1996.
Fill-Out Forms	These are used to improve forms display and entry by creating field labels, grouping related fields into frames, and creating nested forms, as well as providing a new database entry mechanism.
Math	This improvement will allow more sophisticated mathematical functions and graphical creation from tabular data.
Resource Variants	This allows HTML documents that look great on recent browsers to also be backwards-compatible with older browsers.
Toolbars, Frames, and Subsidiary Windows	Users can have specific toolbars and split the screen window into tiled areas, among other things.

5.1.2 *Java*

History and foundation HTML is the most widely used language today for WWW homepage development. HTML is an application of the Standardized General Markup Language (SGML), an ISO standard (No.8879) written in 1986. HTML has been used on the Web since 1990 [2] and has a very active revision history. Even though several browser vendors have added their own unique extensions to HTML the World Wide Web Consortium (W3C) officially maintains HTML specifications. Once HTML specifications are approved by the W3C, they go through the Internet Engineering Task Force (IETF) and become an official RFC for the Internet. To date, HTML has the following W3C version history.

In 1995, Sun Microsystems product engineers assembled together to build advanced software for internetworked consumer electronics. Their original preferred choice of language was C++, but, as they began development, they encountered several problems related specifically to the language. Rather than spend countless hours working around the language problems, they decided to rewrite the language. That rewrite became Java, and its application for Internet programming quickly became apparent.

According to Sun Microsystems, the definition of Java is: A simple, object-oriented, distributed, interpreted, robust, secure, architecture-neutral, portable, high-performance, multithreaded, and dynamic language [3].

Design goals Java's design goals include the following:

- *Simple, object-oriented, and familiar* Little programmer training should be necessary. It has the "look and feel" of C++ and incorporates Object Oriented Development from the ground up.
- *Distributed Java* Built with networking in mind. It contains an extensive library of TCP/IP protocols and WWW functionality, such as URL, HTTP, and FTP.
- *Robust* Java is a strongly typed language since all declarations must be explicitly declared at compile time. This prevents most run-time errors from occurring. Although Java is similar to C++, it does not contain the pointer model that C++ (and C) contains. Thus, there is no pointer arithmetic and no pointer casting. Java has true arrays, which prevent variables from venturing into memory that does not belong to them, as is often the case with C/C++.
- *Secure* Java has built-in security features, which can prevent any unauthorized code from penetrating and creating viruses or corrupting file systems. This is paramount to the creation and sustenance of Java, as it is designed to exist in a distributed environment.
- *Architecture-Neutral* When multi-platform software is released today, it is necessary to create separate versions for each platform (e.g., one for Windows, one for Mac, one for UNIX, one for NT, etc.). Java overcomes this barrier by generating bytecode instructions that are CPU architecture independent. Thus, only one version of Java code needs to be created and it will run on any platform.
- *Portable* Architecture neutrality defines a good portion of portability, but Java offers additional features to portability including strict representation of all data types. Data types are represented the same on every machine (e.g., a *float* is always a 32-bit IEEE 754 floating-point number). The architecture neutral and portable components form the *Java Virtual Machine*. The Java Virtual Machine is based on the POSIX interface specification – an industry standard definition for a portable standard.
- *High performance* The Java compiler creates bytecodes. These bytecodes are translated on the fly by the linker into machine code. According to Sun, this is somewhat like putting the final machine code generator in the dynamic loader [4]. Sun also claims that Java performance is mostly indistinguishable from native C or C++ programs. Additionally, Java comes with an "automatic garbage collector," which runs in the background. Its purpose is to clean up any unused memory.
- *Interpreted Java* Programs are interpreted, which means that Java's interpreter can execute the bytecodes *directly* on any machine that has ported the interpreter and

run-time system. Since Java's linkphase is simple, prototyping is a very attractive piece of functionality.

- *Multi-Threaded* Java has the ability to run multiple threads at once, thus achieving a high degree of interactivity. This is somewhat limited, however, by the platform on which it is running. For example, Windows support for multithreading applications is not as robust as X Windows.
- *Dynamic* Although Java's compiler is very strict during static compile time checking, the run-time version of Java is dynamic in its linking. Classes and code modules are linked only as needed and may be downloaded from across the network. This allows transparent updating of applications as needed.

Status As of January 1996, Java was in its second beta release of version 1.0. Several companies have signed agreements to run Java on their platforms/products (see table 5.5).

Table 5.5 Companies Using Java

Date	Company	Comment
10/30/95	Oracle	Announced license of Java with ORACLE browser for use in accessing software applications through networks
11/8/95	Borland International	Announced license of Java for its Internet tools product
11/8/95	Spyglass Inc.	Announced license of Java for the Mosaic browser
11/10/95	Metroworks	Announced license of Java (for Macintosh platforms)
12/4/95	Netscape	Netscape announced a partnership with Sun to create JavaScript, an "open, cross-platform object scripting language for the creation and customization of applications on enterprise networks and the Internet." They also signed a Java license agreement for Netscape Navigator version 2.0.
12/4/95	Silicon Graphics Inc. (SGI)	Announced license of Java
12/6/95	IBM	Signed Java license agreement
12/6/95	Adobe	Licensing Java for use with Adobe Acrobat Reader and Adobe Page Mill Web Authoring Software
12/7/95	Microsoft	Signed a letter of intent for a Java technology source license
12/21/95	Symantec	Symantec's Espresso software now in synch with the Java Beta2 version
1/9/96	SUN	JavaSoft business unit formed
1/18/96	Precision Systems Inc.	Their next version of UniPort software (v. 7.0) will incorporate Java

With this much support of a technology that is only in its second beta of the first release, it is easy to understand why Sun's CEO stated, "Creating an infrastructure around this revolutionary technology will help us get the power of Java to every software developer for the public Internet and corporate Intranets and develop Java as the Internet programming standard" [5].

5.1.3 Language specifics

Java looks very similar to C++, which makes it familiar to many programmers. Some of the language specifics will look a lot like C++; others are revisions of the C++ language and will be pointed out.

Data Types Everything in Java is an object, with the exception of primitive data types. Table 5.6 lists the primitive data types.

Table 5.6 Primitive Data Types

Data Type	Description
Boolean	This is a departure from traditional C/C++, but Boolean is common in several languages. It takes the value of TRUE or FALSE.
Character	Characters consist of 16-bit Unicode characters. These characters range from 0 to 65,535 to expand the market for global internationalization.
Numeric	Integers consist of 8-bit *byte*, 16-bit *short*, 32-bit *int*, and 64-bit *long*. There is no unsigned data type (as in C/C++). Real numbers are 32-bit *float* and 64-bit *double*.

Arithmetic and relational operators All the normal operators from C/C++ apply in Java. One additional operator was added: the >>> operator to allow an unsigned logical right shift of bits. Also, as with some other high-level languages, the + operator allows string concatenation.

Arrays Arrays in Java are objects with real run-time representation. The user can allocate arrays of any type. Accessing and manipulating arrays is similar to C-style indexing (e.g. `array_name` [3]), but all accesses are checked to make sure they are within the bounds of the array. No C pointer arithmetic is allowed.

Strings Strings are also objects with real run-time representation. There are actually two kinds of string objects: the `StringClass` (for read-only objects) and the `String-Buffer Class` for string objects the user wishes to modify. Strings can be instantiated at declaration time (the compiler understands this). Thus, the line of code:

```
String state = "New Jersey";
```

declares a string *state* and instantiates it to contain the Unicode characters for "New Jersey."

Memory management Java frees the C/C++ programmers from explicitly managing memory. Programmers are not required to allocate and free memory when necessary. There is no `malloc` or `free` commands, as in C. Java's memory management is based on objects and object references. All objects have symbolic object references that refer to them. If the reference is no longer needed, the object becomes a candidate for the Garbage Collector.

Garbage Collector The Garbage Collector runs in the program's background (on one of Java's threads) at a moment when there is infinitesimal CPU downtime. The Garbage Collector gathers up unused memory, compacts it, and frees it. This enables other programs to use this newly available memory.

Features removed from C and C++ The Java designers discovered that one of the inherent troubles in C and C++ was that there were too many ways to do the same thing. Java's goal of *simplicity* attempts to address this by removing redundant or unnecessary items from C and C++. The following is a list of C/C++ language pieces not part of Java:

Table 5.7 C/C+ Language Pieces Not Included with Java

Feature Removed	Description
1. TypeDefs, Defines, PreProcessor	There are no `#define` or `#typedef` commands and no PreProcessor substitutions. Thus, header files are also eliminated.
2. Complex Data Types `Union` and `Structure`	Structures and Unions have been replaced by `classes` with appropriate instance variables.
3. Functions	Real object-oriented programming exceeds functional and procedural paradigms. Anything that can be done with a function can be done by creating a `class` and a `method` for that class.
4. Multiple Inheritance	In C++, objects may inherit characteristics from more than one object (i.e,. a child can have two separate parents). C++ devised a scheme to deal with it. Java just refuses to allow it.
5. Goto Statements	Studies on approximately 100,000 lines of C code found that `goto` statements were used approximately 90 percent of the time to break out of a nested loop[3]. Java has a cleaner implementation called multilevel break and continue.
6. Operator Overloading	C++ allows the programmer to add additional functionality to an operator, based on the arguments passed to the operator—for example, in C++ the + operator defaults to addition of numbers. It can be enhanced to provide other functionality, e.g. string lookup by defining an added class. For example, default +: 3+3 returns 6. After declaring a new class and adding a string lookup parameter to + this, `abcdef+cd` returns 3, the position in the string. Note that I did not show the syntax for defining the new+class in this example, I only pointed out that the + operator can now be used two ways depending on the parameter passed. Java removes this ability as it renders confusion.

Table 5.7 C/C+ Language Pieces Not Included with Java (continued)

Feature Removed	Description

7. Automatic Coercion	Java prohibits C/C++ type operator coercion. The following is a valid example of C code to coerce a float into an int (and hence drop the decimal).
	`int myInt;`
	`double myFloat=2.787;`
	`myInt=myFloat;`
	Here, `myInt` would equal 2.
	Java would cause a compiler error on the above. Type coercions must be done by casting, as the following correct Java example illustrates:
	`int myInt;`
	`double myFloat=2.787;`
	`myInt = (int) myFloat;`

8. Pointers	Java has *no* pointers. Numerous studies show that C type pointers are one of the primary features that allow programmers to inject bugs into their code [3]. Anything that can be done with pointers can be done with Java arrays and objects.

Security in Java As Internet use expands, a great demand is placed on providing highly secure applications, such as electronic software distribution, multimedia content, and electronic commerce in the form of *digital cash*. While security is most often understood at the application boundary layer (i.e., who can access it and who cannot), security is also a consideration at development time, in order to prevent developers from creating subversive code. This is where Java's security features will be implemented. The Java environment starts by assuming that *nothing* can be trusted. This section will discuss memory allocation and layout, the bytecode loader and verification process, and security in the java networking package.

Memory allocation and layout within Java occurs on the host machine at run time. Java programs are prevented from *drifting* into the memory area of another program or another Java program. Java has no pointers, so hard-coded calls to memory are not possible. Additionally, Java programs are only allocated a specific section of free memory.

When a Java program is initiated, bytecodes are sent from the server machine to the client machine. The Java run-time system, which resides on the client machine, explicitly distrusts the bytecodes that come across the network and, prior to program execution, loads the bytecodes into a through a byte code verification process. Figure 5.1 illustrates the Java program cycle.

The Java run-time system does not care where the bytecode information comes from. It could be on the local client machine or it could come from a server in another country. It applies the same verification process to all bytecodes received. This effectively separates the two environments (compile and runtime) from each other even if they are on the same physical file system. The ByteCode verification process checks for the following things to insure a secure run-time application:

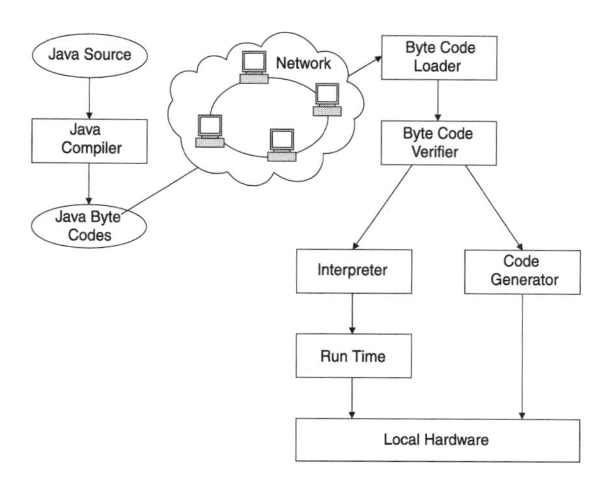

Figure 5.1 The Java program cycle

- The bytecode format is correct.
- The bytecode does not forge any pointers.
- The bytecode does not violate any access restrictions.
- The bytecode accesses objects as they are.
- The bytecode will not cause any stack overflows or underflows.
- The bytecode parameters are all correct.
- The bytecode contains legal objects with scope of private, public, or protected.

Although this verification process appears to be time consuming, the upfront validation improves run-time efficiency and allows the program to be interpreted fully, without cause for worry, since the interpreter does not have to check for any errors. All reliability checking is done in advance.

Security in the Java networking package allows the user to configure the wasy in which Internet protocols (FTP, HTTP, etc.) may interface with the Java program. The package can be set at the following restrictive levels:

- Disallow all network access (local application only).
- Allow network access only to the hosts from which the code was imported.
- Allow network access only outside the firewall, if the code came from outside the firewall.
- Allow all network access.

5.1.4 *Applications using Java*

Sun Microsystems has created a browser called the HotJava browser. This browser is written in the Java language. This browser differs from standard hypertext browsers, such as Netscape or Mosaic, because it is incredibly dynamic.

When static browsers come across an object (e.g., a GIF image file), they know how to deal with it, because it is part of the configuration of the browser. However, the user must maintain this configuration and if the browser comes across an object with which it does not know how to deal, the browser has no way to deal with it directly and the user must make a decision regarding how to react to this situation. HotJava has no idea internally how to handle a GIF image; rather, during runtime, it pulls in Java code that knows how to respond to this object. Figure 5.2 illustrates the HotJava browser interfacing with an object:

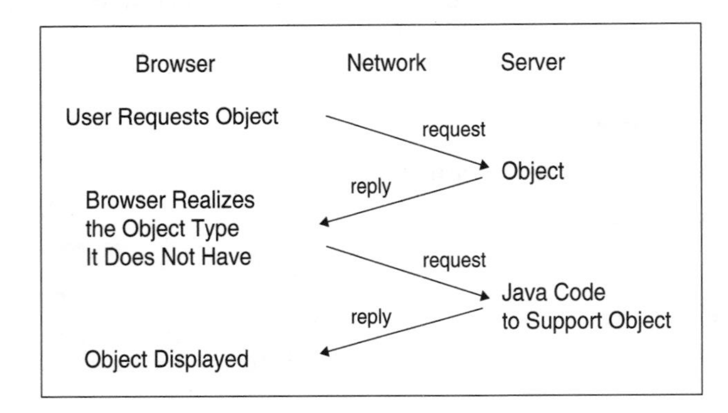

Figure 5.2 The HotJava browser interacting with an object

The same object-oriented technique applies to WWW protocols. This allows HotJava to be very scalable for future proprietary security protocols. For example, if a user wants to access two different WWW sites through HotJava and each web site is running a proprietary network protocol, the browser simply downloads the Java bytecode for the proprietary protocol and establishes the connection. No configuration to the browser is necessary.

Other WWW browsers are implementing Java into their architecture. Among these is Netscape, which formed a partnership with Sun to integrate the Java run-time interpreter into its version 2.0 browser called "Netscape Navigator". Netscape is the first browser vendor to partner with Java. Other companies licensing Java's interpreter include Oracle with its PowerBrowser and Spyglass with its Mosaic Browser [6]. These browsers interact with Java applets—a java code object that is already compiled and in bytecode format ready to be transported across the network and executed on the local machine.

5.1.5 Future of Java

Journalists and some major industry players are hailing Java as the "future internet operating system" [7]. Sun has realized the following additions for 1997:

- Support for other programming languages. Sun wants to support other languages besides Java to run on top of the Java virtual machine. Languages include Ada, Dylan, Scheme, Smalltalk, and MS Visual Basic. Additionally, James Gosling, the Sun scientist who invented Java stated, "Negotiations with other language vendors are happening all over the place" [7].
- Create a compiler that turns these non-Java languages into ByteCodes that can be loaded into the Java interpreter. This would allow several different types of programmers to create Java applets capable of running on the Web.
- Uniform security policy—Sun's plans for security include two parts: A Java applet would not be able to read or write to/from any local file, and a Java applet could only make network connections back to the host from which it came. Sun encourages browser vendors to allow optional settings that would allow more flexibility. After downloading the applet, Sun's HotJava browser currently allows four security levels (most-to-least secure):
  - The applet may not communicate with anything.
  - The applet may only communicate from the server it came from.
  - The applet may only communicate with servers inside the firewall.
  - The applet may communicate with any server (unrestricted).

Sun also plans to add secure digital-signature authentication when downloading applets.

- A higher-speed compiler—Sun intends to create a just-in-time compiler with Java, so that Java code can take advantage of local compiler efficiencies and optimizations (e.g., a C++ compiler). Sun also expects to exceed C++'s compiler [7].
- A Java OCX—Sun plans to release Object Linking and Embedding Control eXtension (OCX) for Windows. This will enable Windows applications to access Java applets directly, instead of through a browser.
- Java Multimedia and Graphics Classes—Sun plans to release new multimedia and graphics classes with partners Macromedia and Silicon Graphics.

5.1.6 VRML

History and Foundation Development of the Internet can be summarized by three distinct phases. First there was the full-scale development of the communication infrastructure known as TCP/IP. This allowed a layer of abstraction between the user and the data. This phase, however, was basically a closed environment to scientists, net-surfers, and those versed in TCP/IP and the basic structure of the Internet. Users in this phase needed to have a cognitive map of the Internet in order to access desired data.

The second phase was the development of the World Wide Web. The Web added an additional layer of abstraction that made the Internet extremely accessible for people who had little knowledge of TCP/IP. Data and resources could be located using an address scheme known as the Universal Resource Locator (URL). It also provided hypermedia links to other addresses. WWW development vastly improved the user interface to the Internet and is the main reason more and more people are accessing the Web today. This phase, however, is two-dimensional and media-based. Users have a much easier time navigating the Internet than before, but there is no intuitive, perceptual means of abstracting the data they need. This phase uses the HTML language.

The third phase presents a way for users to perceptually navigate the Web through a sensual interface. The Virtual Reality Modeling Language (VRML) is the base language for this phase. VRML puts the user at the center of the Internet and lets the user order the Internet universe in his or her own way. The sensual, interactive interface allows the user to use *cognitive perception* to find the needed data. The user will not be required to recall from memory a specific location nor follow a string of addresses to find the needed data [8].

VRML is still a very young technology. It was conceived at the first annual WWW conference in Geneva, Switzerland, 1994. A small group convened to discuss three-dimensional tools that would interact with the Web. The group decided that the tools should have a common language, and they initially dubbed it Virtual Reality *Markup* Language as an analog to HTML. However, due to the graphical distinctions between 3-D Virtual Reality and 2-D markup, the group later agreed to change the name to virtual reality *Modeling* Language to more accurately describe three-dimensional scene modeling. The first VRML draft was completed on November 2, 1994. Two additional revisions were made and the third draft (dated May 26, 1995) was accepted as version 1.0 of VRML. VRML is designed with the following requirements:

- Platform independence
- Extensibility
- Ability to work well over low-bandwidth connections [8].

These requirements are also similar to HTML. Interestingly enough, version 1.0 of VRML includes no interactive features. The basis for this decision was "a practical decision intended to streamline design and implementation" [8]. Indeed, interactive behavior in a virtual reality for a Web browser does not exist yet, based on the three requirements listed above (e.g., none are platform independent), and the VRML group would have needed to build one from scratch—thus delaying the development of the language.

Applications Although VRML is just in an infancy stage, there is already an active application on the Web called "Virtual Vegas." It is a crude 3-D implementation but nonetheless an actual application. No real money is exchanged (paper or electronic). When the user signs up, he or she receives credits to play the games. Games include all casino games, such as blackjack, poker, slot machines and so forth. High winners are given a prize by the sponsors of the page.

Language specifications VRML consists of nodes, which are essentially any type of object (e.g., 3-D objects or JPEG images). Thus, VRML is object-oriented (or node-oriented). As with objected-oriented languages, nodes can have children and may be reused or modified through instancing. Unlike typical object-oriented languages however, VRML creates a *scene graph* or a scene in which the nodes are hierarchically organized (most object-oriented languages employ random organization to objects). This hierarchical arrangement dictates the sequence when the nodes are drawn within the scene. This lets programmers control the appearance and layout of the scene.

Nodes have the following characteristics:

- The type of object (e.g., a cube, sphere, transformation, JPEG image)
- Parameters to distinguish it from similar nodes (e.g., one cube may have different height than another)
- A name to identify the node. This allows the node to be referred from another scene. A name is not required and several nodes can have the same name
- Child nodes [8].

VRML scenes exist within a three-dimensional coordinate system. The Cartesian coordinates consist of an *x, y, z axis* where x is horizontal, y is vertical and z represents the depth (see figure 5.3).

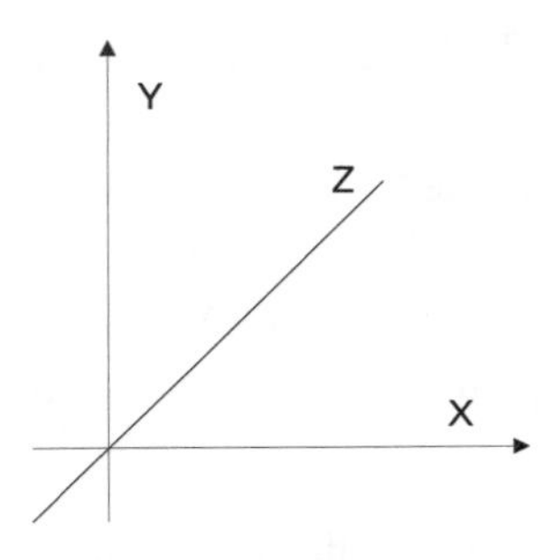

Figure 5.3 x, y z coordinate plane in VRML on a 3-D monitor

The default coordinates (0,0,0) are the bottom left corner of a two-dimensional workstation As *z* decreases, the object retreats into the depth of the scene. Conversely, as *z* increases, it approaches the user. The standard unit for distances and lengths is meters; for degrees, it is radians—for example, to rotate an object 180 degrees about its *x* axis, the programmer would specify that it be rotated about the *x* axis 1p radians, or 3.14.

VRML currently consists of 36 types of nodes. Most nodes can be classified as either *shape, property or group*. The following is a list of VRML nodes, their type and a short description:

Nodes contain fields within them. There are two general classes of fields; fields that contain just a single value and fields that can contain more than one value (like an array). The following is a list of fields, their type and a short description:

Table 5.8 VRML Nodes

Node	Type	Description
ASCIIText	Shape	Plain ASCII Text
Cone	Shape	A simple cone
Coordinate3	Property	This defines a set of 3-D coordinates to be the *current* coordinates. It is often used with Indexed-Face-Set, IndexedLineSet and PointSet.
Cube	Shape	A simple cube
Cylinder	Shape	A simple cylinder
Directional Light	Property	This offers more control over a scene's light display. This light runs parallel with the directional light field vector.
FontStyle	Property	Defines the font style for all subsequent ASCIIText
Group	Group	This is the base class of group nodes; it contains an ordered list of child nodes. It is considered a logical association of nodes that are related in some way.
IndexedFaceSet	Shape	The most commonly occurring geometric shape in generated scenes, this represents a 3-D shape built from polygonal faces.
IndexedLineSet	Shape	This represents a 3-D shape formed by constructing polylines from vertices located at the current coordinates.
Info	Property	Defines an information node that would hold application-specific data about the scene (e.g., copyright information).
Level of Detail (LoD)	Property	This allows various levels of detail to be portrayed, based on the distance away from the world-eye—for example, if a painting were on the back wall of a scene, it would appear with minor detail from a distance and increase in detail as the user moved closer.
Material	Property	Defines current surface materials

Table 5.8 VRML Nodes (continued)

Node	Type	Description
Material Binding	Property	Specifics how current materials are bound to shapes
MatrixTransform	Property	Defines a geometric 3-D shape into a 4x4 matrix through transformation
Normal	Property	Defines a set of 3-D vectors that will be used to reflect light perpendicularly when light is drawn
Normal Binding	Property	This specifies how current normals are bound to shapes in a scene. Each shape may interpret bindings differently.
Orthographic Camera	Property	Translates a 3-D scene onto a 2-D plane without reducing object size. This gives the impression that a scene has no depth—that is, objects are not smaller from far away; they are the same size from every perspective.
Perspective Camera	Property	This translates a 3-D scene onto a 2-D plane but diminishes objects as they recede into the background. This gives the impression of depth.
Point Light	Property	Light emanates from a single point in infinite directions. This is somewhat like a plain light bulb without a lamp shade.
PointSet	Shape	Represents a set of points at the current coordinates
Rotation	Property	Defines a 3-D rotation about an arbitrary axis through the node's origin
Scale	Property	Defines a 3-D scaling about the node's origin (e.g., a scale of 2,2,2 would increase the size of the node 200 percent in each *x,y,z* direction).
ShapeHints	Property	This indicates if IndexFaceSets are solid, or contain ordered vertices, or contain convex shapes. This can aid scene-rendering optimization.
Sphere	Shape	A simple sphere
Spot Light	Property	Light emanates from one place but into a conical shape disbursement.
Switch	Group	A control structure that can be instructed to traverse one, none, or all of its child nodes
Texture2	Property	Defines a texture that may be applied to subsequent shapes as they are rendered
Texture2Transform	Property	Defines a way for the user to alter the texture on a shape from a specific center point
TextureCoordinate2	Property	This defines a set of 2-D coordinates used to map textures to vertices. It replaces current texture coordinates in the rendering state.
Transform	Property	Defines a geometric 3-D shape transformation to a point, around a point, or to a translation
Translation	Property	Defines a geometric 3-D movement linearly only
WWWAnchor	Group	This node is similar to the <A> anchor tag in HTML. When a WWWAnchor node is selected, a call is made to the Universal Resource Locator and a new VRML scene is loaded.
WWWInline	No Category	WWWInline could make or break VRML. It allows different objects in a scene to be stored in separate files and loaded only if requested. This potentially prevents the viewer from loading huge quantities of data over a slow connection. This node type also allows these "files" to be classified into library types, so that instead of loading from a distant server, the child object can be loaded from a local server or even the user's machine itself.

Nodes can contain many fields. There are two general classes of field: fields that contain just a single value and fields that can contain more than one value (like an array). Table 5.9 is a list of fields, their type, and a short description

Table 5.9 Fields

Field	Type	Description
SFBitMask	Single	This is a bit mask that specifies actions for different parts of a node—for example, when drawing (or rendering) a cone, there are three parts that can be drawn: all, bottom, or sides. The bit mask can be set for each of the three parts.
SFBool	Single	A true or false variable
SFColor	Single	This is a color variable consisting of measured shades of red,green or blue. Each color is concentrated with intensity from 0.0 to 1.0 and mixed with the other two colors.
SFEnum	Single	Enumerated data type for mnemonic labels
SFFloat	Single	Single precision floating-point number
SFImage	Single	Bitmap image stored by pixel: Image is based on the *x,y,z* coordinate axis and is offset from 0,0,0.
SFLong	Single	32-bit integer
SFMatrix	Single	This is a transformation matrix coded as 16 linear values. A transformation matrix describes how an object is to be drawn (i.e,. rotation around axis and position relative to *x,y,z*).
SFRotation	Single	This is an arbitrary rotation of *x,y,z* in radians—for example, 0,0,3.14 represents a rotation of the object around its *x* axis, 3.14 radians or 1/2 turn.
SFString	Single	A character string
SFVec2f	Single	A 2-D vector written as a pair of SFFloats
SFVec3f	Single	A 3-D vector written as a trio of SFFloats
MFColor		An array of SFColor
MFLong		An array of SFLong
MFVec2f		An array of SFVec2f
MFVec3f		An array of SFVec3f.

Browsers Since there are few users of VRML applications, there are very few browsers available. The following lists some browsers for varied environments.

Windows platforms

- VRWeb—made by a consortium from the Institute for Information Processing and Computer-Supported New Media, the National Center for Supercomputing Applications (NCSA), and the Gopher development team from the University of Minnesota.

- WebSpace—made by an alliance between Silicon Graphics Inc. and Template Graphics Software.
- WorldView—made by InterVista Software [9].

Macintosh Platforms

- Whurlwind—the first working VRML browser on a Mac, written by William Enright and Jon Louch.

X Windows (UNIX) platforms

- VRWEB
- WebSpace
- WebView

These three products were written by the same authors who wrote for MS Windows.

Limitations and the future of VRML Currently, with only VRML 1.0 available, very few applications are using VRML. The sheer amount of computation involved in 3-D graphic generators poses limitations for to-dos desktop workstations. Improving technology, however, could offer some of the following uses for VRML development on the WWW:

- Architecture—homes could be displayed on the Web for prospective buyers. Ambient lighting, furniture arrangements, and even summer versus winter light could be included as variables in the description.
- Virtual storefronts—online shopping could really expand through VRML. Users could enter their measurements, for example, and see several articles of clothing in several different colors. The user could then select the piece he or she wants, determine if it is in stock and order it.
- Medical research—students could perform an autopsy or brain surgery on a virtual patient.
- Entertainment—there is already a Virtual Vegas site offering 3-D virtual gambling. The other entertainment possibilities are endless. One entertainment industry likely to do massive business would be sex-related material. [9]

5.2 Web development editors

There are a number of editors on the market today for Web development. Since HTML is the predominant language, HTML editors are the predominant editors. It should be noted that HTML can be created with just a simple text editor (e.g., *vi* in *UNIX*, or *Edit* in DOS). However, HTML–specific editors offer the developer better protection from errors and often allow the developer to create documents in a WYSIWYG environment. HTML editors also come in flavors that support HTML (pre-2.0), HTML2.0, HTML3.0, and HTML3.0, extensions (i.e., extensions by specific browsers but not currently approved by the World Wide Web Consortium).

HTML editors also are available for several platforms such as UNIX, DOS, Windows, Windows 95, Macintosh, and X Windows. On one text search of HTML editors, the Yahoo! search utility retrieved 74 different editors. Prices range from $50 to $250. There were also several free HTML editors. HoTMetal Pro by SoftQuad is a very popular HTML editor used by AT&T's corporate Web page development staff. Some of its features include the following:

- Word processing conversion into HTML
- WYSIWYG table editing
- Automatic rules checking to produce error-free HTML
- Free upgrades to HTML extensions
- Easy-to-use forms creation and image maps
- Preview of your work with built-in browser (or access to any browser)

5.3 Validation tools

There are several validation tools that assist in the development of an HTML Web page. This section will discuss two HTML validation tools called Weblint and HtmlChek as well as a hyperlink integrity tool known as MomSpider.

5.3.1 Weblint

A commonly used HTML code checker today is *Weblint,* written by Neil Bowers of Khoral Research Inc. The extension "lint" is most likely borrowed from the UNIX/C *lint* program, which performs additional checks against C code that a compiler does not perform. Weblint is written in the perl language and checks an HTML file for the following conditions:

- Verify the basic structure
- Check for unknown elements or element attributes
- Check the context
- Check to see if "TITLE" is in the "HEAD" element

- Check to see if "IMG" elements have "ALT" text attribute in case the client machine is incapable of displaying the image
- Check for illegally nested elements
- Check for mismatched tags (e.g. `<H1>` ... `</H3>`, should be `<H1>` ... `</H1>`)
- Check for unclosed elements (e.g. `<H1>` ... should be `<H1>` ... `</H1>`)
- Check to see if one-time occurring elements occur more than once
- Check for any obsolete elements
- Check for an odd number of quotes within a tag
- Check to see that the order of the headings is correct
- Check to see if there are any markup/hyperlink references embedded within comments (some browsers are confused by this)
- Check to see that attributes are located where they're supposed to be
- Check HTML3 elements such as TABLE, MATH, and FIG Checks for unclosed comments.

Weblint is free for FTP downloading.

5.3.2 HtmlChek

HtmlChek is similar to Weblint in that it validates HTML code. HtmlChek focuses mostly on syntax and is capable of checking both HTML2.0 and HTML3.0 syntax programs. HtmlChek was written by the University of Texas, and is free for FTP download.

5.3.3 MomSpider

The MultiOwner Maintenance Spider is a robot program that performs automatic validation on hyperlinks in a document. MomSpider can traverse a list of Web sites (by owner, site, or document tree), check each Website for problems that require the document owner's attention and build a special index listing the problems. This frees a document owner from manually reviewing his or her Webpage to see if all links are correct, since MomSpider runs automatically and displays all current problems. MomSpider is available in the UNIX environment and is written in perl.

In addition to running automatically, MomSpider is also more efficient than a mere mortal running a validation process. When manually validating a link, a human issues the http: protocol which calls a Get method (i.e., pulls document content data across the network). MomSpider uses the Head request method. This method attempts to access the Website and, if successful, extracts the Last Modified Date and Expected Expiration Date. These data would then be written to the special index. Since no content data comes across the network, MomSpider is much more efficient. With the knowledge of a remote site's expiration date, a programmer could update his or her page at the appropriate time.

MomSpider does not correct any link mistakes. It simply writes them to an index file. Correction remains the responsibility of the programmer.

MomSpider was written by Roy Fielding of the University of California (Irvine) and is free for FTP downloading.

5.4 HTML converters

In a recent search of the Web using InfoSeek's search engine, there were approximately 100 HTML converters (both to and from HTML). Table 5.10. is a sample listing of some converters with a short description of what the program does.

Table 5.10 HTML Converters

Converter	Description
WP2X	Converts WordPerfect to HTML
rtftohtml	Converts RTF to HTML (for Mac and UNIX formats only)
MSWord Macro	An MSWord macro that converts Word files to HTML
Excel Macro	An Excel 5.0 macro that converts Excel tables to HTML tables
CU_HTML.DOT	An MSWord 2.0 template that allows HTML files to be developed
WebMaker	Converts FrameMaker documents to HTML
pm2html	Converts PageMaker documents to HTML
AmiWeb	Converts Ami Proc documents to HTML

5.5 References

1 D. Kinsky, *Software Tools in Support of WWW Development*, (a Class project, TM601, Telecommunication Technology, Stevens Institute of Technology, April 1996.)

2 T. Berners-Lee, *HTML and Style Sheets a Working Draft*, located at *http://www.w3.org/Hypertext/www/tr/wd-style*.

3 S. Hassinger and M. Erwin, *Internet World: a 60 Minute Guide to VRML*, IDG books, Forest City, CA, 1995.

4 R. Karpinski, *Java: Sun Opens Door to Java Development*, Communications Week, January 8, 1996.

5 J. Levitt, *Java Makes the Web Perform*, Information Week, (January 1, 1996 p.55-57).

6 M. Pesce, A. Parisi; B. Gavin, *The Virtual Reality Modeling Language*, located at: *http://vrml.wired.com/vrnl.tech/vrml10-3.html.*

7 Sun Microsystems, *Introduction to Java*, located at *http://java.sun.com/whitepaper/java-whitepaper-3.html.*

8 Sun Microsystems, *Overview of the Java Language*, a *White Paper,* located at *http://java.sun.com/1.0alpha3/doc/overview/java/index.html.*

9 Sun Microsystems, *Sun Microsystems Announces Formation of JavaSoft,* located at *http://java.sun.com.pr-javasoft.html.*

PART 2

Introduction to ATM and Quality of Service

CHAPTER 6

An Overview of ATM

This chapter provides, for convenience, a short tutorial on ATM. It is assumed that the reader already has some working knowledge of the field, and so the description is limited in scope. If more background on ATM is needed, the reader may refer to a number of texts, including reference [1]. More detailed background features are covered in the following chapters.

6.1 Introduction

ATM is an evolutionary technology. It takes full advantage of the large-capacity potential of single-mode fiber-optic media, while adjusting to the limitations of today's infrastructure, which will remain in place for a number of years. Specifically, interworking with legacy networks and protocols is a feature that has been built into ATM. At the local area level, ATM uses advanced encoding schemes to support 25, 100, and 155 Mbps on twisted-pair and multimode fiber media.

ATM is an enabling technology, delivering a capacity level that could change the nature of networking both at the local and at the wide area level. Since it has been developed over a number of years, ATM is the next step in a progression. Private networks have evolved over the past quarter century. Successive architectures—all still in use—include X.25, Ethernet, Token Ring, FDDI, and frame relay. ATM technology is capable of overcoming its antecedents' shortcomings in bandwidth, scalability, traffic-handling capabilities (this is called traffic management), and suitability for combined data, image, and voice traffic.

ATM technology offers the following benefits, among others:

1. It supports high-speed communication (DS3, OC-3, and higher in the future).
2. It is a scalable technology, so higher speeds can be supported without having to obsolete the underlying technology infrastructure.
3. It supports both WAN and LAN systems, enabling the user to deploy a single technology for all the enterprise-wide needs.
4. It supports multimedia communications.
5. It supports multipoint communications.
6. It supports quality-of-service contracts.
7. It can be used as a multiservice platform to support a variety of legacy services, such as LAN bridging, LAN Emulation, frame relay, and circuit emulation.
8. Its technology is cost-effective (because of the overbooking of user bandwidth), reliable, and well supported by the industry.

In a smaller sense, ATM is a cell-based data link layer (DLL) protocol. A data link layer protocol has the job of moving information across a single link (in a reliable manner), and using well-defined frames (here, cells). In this view, ATM only fulfills the lower portion of the data link layer. The data link layer protocol relies on a physical layer (PHY) protocol, which can operate in any media, including DS3, OC-3, TAXI/FCS multimode fiber, and UTP (CAT5). Protocol-wise, ATM and the PHY, taken together, corresponds to a MAC layer of a LAN (except that the former is connected to the DLL). In general, there is a need to fill out the DLL, before a network-layer protocol (e.g., IP, IPX) can be carried. This is done using an AAL set of protocols. LANE and MPOA also address these interworking issues. It should be further noted that the protocol model has three stacks: the user plane, the control plane, and the management plane. ATM can be used in all their planes as the common link protocol, although the upper layers would be different.

In a more colloquial sense, ATM is the entire technology that supports fast packet broadband communication. It is a statistical multiplexing technology, where the industry has agreed on how to support open multiplexing. With this technology, providers can overbook user bandwidth, thereby reducing network facilities; this in turn reduces provider costs, which, in the final analysis, can reduce the cost to the user.

6.2 The emergence of ATM

Computing has, for the past 12 or more years, followed Moore's Law: the density (and, hence, the potential performance) of silicon-based microprocessors doubles every 18 months. This is about an order-of-magnitude improvement every five years. This observation, named after its originator, Intel cofounder Gordon Moore, is governed by technology. By contrast, networking technology has advanced more slowly than computer technology, as measured by, say, user-available bandwidth. It was shown in reference [4], in what might be called an equivalent "law," that bandwidth of a WAN has increased by an order of magnitude every 20 years (this "law" is based on an analysis covering 60 years). Raw bandwidth growth has been governed principally by technological factors. However, the traditional gap between computers and networks has also been influenced by the lack and ineffectiveness of a complete set of standards, their slow evolution, and their often-cautious adoption. As a result, ten-year-old networks are commonplace, and many are likely to remain in service for some time. Network software has, on the whole, advanced even slower than computer software. Networking standards are needed to make it possible to connect systems of numerous manufacturers and facilitate interoperability of networks with different specifications.

A network is also limited by the capacity of the transmission medium through which data and other signals are transferred. These physical channels have evolved from twisted-pair wire, intended initially only for voice traffic, through more advanced uses of copper, to today's fiber-optic facilities. Electronic switches have gradually displaced electro-mechanical equipment. These diverse physical facilities of public networks, spread all over the world, represent a large investment. Telecommunication companies are accustomed to multidecade system lifetimes. Therefore, there is a tendency to retain embedded technology for a significant amount of time. Nonetheless, ATM is driven by advances in fiber-optics. The capacity and overall cost/effectiveness of fiber are attractive and are giving some impetus to the transition. Fiber is already predominant for major long-distance channels in the United States.

The ATM standards were originally based on Broadband Integrated Services Digital Network (B-ISDN) standards, which in turn evolved from ISDN standards. For LAN and campus environments, a number of newly developed standards have also evolved recently.

Many private networks are also moving to improve backbone as well as desktop capacity. Refinements in copper-wire technologies and the cost of replacing existing investments will, however, delay the introduction of fiber for private LANs at the desktop level. Recent developments are extending the practical lifetime of existing media in private networks:

- Improvements in carrying capacity and error reduction in copper-based LAN runs
- Introduction of switching into the traditional shared media networks, such as Ethernet, Token Ring, and FDDI
- Adoption and near standardization of data compression techniques, which now reduce the bandwidth demands of images and video, as well as text, making 10Base-T and 100Base-T LANs at least marginally capable of supporting these media

For a while, at least, all these network architectures will continue to play roles in private networks. Increasingly, however, ATM will displace these architectures or be utilized as an intermediate layer, supporting other protocols such as TCP/IP. ATM offers the following features:

- It is suitable for both long distance and local networks.
- It adapts to a wide range of physical media.
- It accommodates voice, data, images, and video.

- It is capable of supporting quality of service, congestion control, and other network management features.
- It can adjust to bursty traffic and avoid congestion.
- It scales to much higher bandwidths than frame relay or other competitive architectures.

An ATM platform can be used to support a variety of legacy services, such as LAN bridging, LAN Emulation, frame relay, circuit emulation, and so forth. The switch and/or Customer Premises Equipment (CPE) concentrator provide interworking (protocol conversion) functions, which enable the legacy frames (and/or bit streams) to be mapped into ATM cells. This is done by using appropriate AAL protocols.

6.2.1 ATM's evolution

ATM has evolved from previous network architectures. Much has been done recently to upgrade LAN and WAN technologies. Both were originally based on copper wire, whether unshielded twisted pair, shielded twisted pair, or coaxial. The quality and carrying capacity of copper media have been improved steadily, and these solutions remain useful for smaller LANs and WANs (particularly the access portion to the WAN) that do not generate high volumes of traffic. In a wider perspective, however, both LANs and WANs are approaching their limits. Ethernet, Token Ring, and comparable networks are the foundation of today's routine network functions: sharing files, the client/server philosophy, and clustering workstations or other computers together to form a single distributed computational system.

FDDI, a technology based on fiber, went into use in the late 1980s. It is also based on a Token Ring philosophy. FDDI's nominal transmission limit, 100 Mbps, is ten times greater than the capacity ceiling of Ethernet. FDDI makes it possible for network managers to control traffic flow, but it does not carry this capability very far. In the early 1990s, it became clear that a faster LAN technology would be necessary to keep pace with in-building bandwidth demands. However, FDDI was deemed too costly and complex. Two extender technologies were proposed, each of which was based on the idea that the fundamental philosophy behind Ethernet could be extended to provide more bandwidth. These technologies were thus known as Fast Ethernet. As discussed in chapter 2, 100Base-T was compatible with Ethernet (just at 100 Mbps). The other, 100Base-VG, had additional functionality. These high-speed LANs serve a niche market; however, each is little more than an extension of legacy technologies. They do not have the capability to be an enabling technology, because many of the intrinsic limitations of shared media remain.

Development of ATM technology officially started in January 1985, although trial work started a few years earlier. By 1989, many of the standards required to support WAN data communication were available, and equipment started to appear. Additional standards and equipment have appeared ever since. For WAN data applications, ATM technology is well into the third generation of equipment,.

What makes the difference between ATM and the solutions from which it has evolved? The following paragraphs will discuss this.

ATM offers architected features such as quality of service, statistical multiplexing, and traffic loss prioritization. ATM per se does not support per-cell treatment priority from a performance point of view (there is a loss priority, but this is not directly related to expediting cells through the switch or rendering special treatment). To compensate for this, implementers have developed an "external" mechanism of providing per-session "traffic" priority discrimination. In a Permanent Virtual Circuit (PVC) environment, the cells belonging to a session are implicitly assigned a priority at the switch as the cell traverses the switch (this supplementary information is, however, removed as the cell exits the switch); routing cells (based on the supplementary information) to different buffer pools. This implicit mechanism allows different traffic flow to receive different switching treatment (e.g., constant bit rate, variable bit rate, etc.). For switched permanent connections, the treatment of the block of cells associated with a connection is based on the kind of service class called for in the setup message sent by the user for that connection, at the start of the session. Again, the switch will have to use extensions of ATM to service the different requirements; in addition, the priority is not on a per-cell basis. Hence, the issue of "priorities in ATM" has to be carefully worded: one can say that mechanisms exist to support a kind of connection-level grade of service, which gives the appearance of supporting, from a performance point of view, a weak kind of traffic priority. ATM supports a connection-oriented service, rather than a connectionless service. With its "guaranteed" quality of service, it is better suited to carry video and multimedia.

ATM's bandwidth scalability is capable of accommodating, with appropriate hardware upgrades, the expected growth in end user bandwidth demand. This expected increase is driven by the rapid increases in performance and output of the computers served by the networks, and by the growth in numbers of users, as desktop systems are becoming almost universal. ATM's basic philosophy of scalability provides technological longevity.

Desktop workstations based on the most powerful Reduced Instruction Set (RISC) microprocessors—for example, Digital Equipment's Alpha—can already swamp an FDDI connection when used for I/O intensive functions. PCs will reach similar output in a generation or two. As desktop users have deployed applications such as Microsoft Windows and Lotus Notes with appended voice and video, demand for network capacity has accelerated. ATM will be able to better support these and similar applications than existing communication services.

It is essential that bandwidth be allocated to users and applications intelligently and efficiently. This feature cannot be added as an afterthought; it must be part of the original architecture—one of ATM's strongest features.

In addition to commoditizing the physical layers, the much simplified infrastructure made possible by ATM's basic philosophy permits a much longer productive life cycle for the cabling plant and networking hardware than for other (LAN) technologies. The cost and inconvenience of performance upgrades are reduced—for example, a change from ordinary Ethernet to Fast Ethernet requires a comprehensive replacement of adapters and other hardware. The ATM/SONET interface permits increases in speed with fewer, less-costly adjustments. However, it must be understood that if the user purchased a 51.84 Mbps card, this card will likely not support 155 Mbps or 622 Mbps; a new card would be required. The advantage of this approach is that the entry-level price is small, since the user is not buying features that are not needed. On the other hand, the user may choose to purchase a more powerful (but more expensive) card on day one, which can, by design, support 51.84, 155, and 622 rates immediately. The user would initially only use this card at the lower rate. As needs increase, the user can then, without any hardware changes, upgrade the speed. It should be clear, though, that a tradeoff is required—nothing is ever free.

ATM strikes a balance between two related issues in network design: accommodation of traffic bursts and control of congestion caused by competition for resources among candidates for transmission, particularly in the presence of bursts. Ethernet in the LAN and frame relay in the WAN cope with surges in network traffic, but can become congested if too many computers on the network want to transmit at the same time. Token Ring has a monitor that controls access, but it can backlog computers for unacceptable lengths of time, waiting until a token lets them on the network. However, in Token Ring the delay is deterministic and bounded, while in Ethernet it is stochastic and (theoretically) unbounded. ATM, in contrast, supports a form of bandwidth on demand, up to the maximum speed of the access line and/or switch, and consequently can deal more easily with bursts. It should be noted, however, that ATM has a limited bandwidth equal (at the switch level) to the bus speed—say 2 Gbps; also, there is a certain number of buffers—say 64,000 per line card. As soon as the user puts in more than a certain well-calculable number of inputs and arrivals, the user can become *dead in the water,* just as in frame relay and Ethernet. In addition, there are constraints based on both the access speed of the link (user/line side) and the speed of the trunk.

As with any other technology operating at the data link layer, ATM is independent of upper layer protocols (so are, in principle, Ethernet, Token Ring, etc.). This means that ATM will carry any Protocol Data Unit (PDU) that is handed down to it through the syntax of the ATM Service Access Point (SAP). From ATM's perspective, it is immaterial what is contained in the cell: IP, IPX, SNA, protocol control information, or payload. From the upper-layer protocol's point-of-view, however, there has to be protocol compatibility matching: the protocol's PDUs must conform to the ATM's SAP; hence, the upper-layer protocol must know how to hand off its PDUs to ATM. Having noted this tautological observation, there is a desire to retain existing upper-layer protocol applications unchanged. Hence, the developers have established appropriate ATM adaptation layers to accommodate the interworking functions. AALs reside above the ATM layer and below the network layer. Note that ATM is a connection-oriented technology. This means that connections must be established (at the ATM layer) before information can be exchanged. There is also a desire to develop virtual LANs (VLANs), where the logical community definition is independent of the physical location of the user. A number of products already exist to support VLANs, but these are vendor-specific. Many hope that ATM (in particular LAN Emulation technology) will become a vehicle to deliver vendor-independent VLANs.

In ATM, the view of the network is nearly the same, whether it is a LAN or WAN, public or private. Thus, although ATM is an evolutionary step forward from earlier networking technologies, it is a major step.

Legacy networks are supported. In an interworking mechanism known as LANE, Ethernet and Token Ring frames would traverse an ATM network (in a segmented fashion) and be delivered transparently to a similar legacy network on the receiving end. Furthermore, a user on an ATM device can send information to an Ethernet or Token Ring architecture, without passing through successive stages of large-scale reinvestment. Another migration, discussed in following chapters, is MPOA. To dispel some misconceptions, it must be noted, however, that the prospective user has to make an immediate investment to acquire LANE technology in order to support some of the ATM functions at this time. The transition is then from Ethernet/Token Ring to LANE to ATM. Another approach would be for the user to save the funds invested in this partial migration to ATM, in order to be better equipped to then make the direct migration from Ethernet/Token Ring to ATM at a later date. The LANE solution requires investments just to support today's relatively low bandwidth requirements; however, later it could be difficult to scale up to meeting the challenges of evolving network traffic patterns. Traffic concentration using ATM's higher capacity could overload tributary Token Ring and Ethernet networks by dumping bursts of information that choke the receiving network. The need to protect investments must be traded off against the risks of poor overall performance. In addition, solutions must be found for problems of addressing brought about by connections with legacy networks. Even within the ATM domain, problems raised by address resolution and flow control have not been solved completely. It is also desirable to integrate frame relay into ATM. Alternatively, ATM circuits might have to be utilized at lower speeds than SONET speeds in order to accommodate legacy traffic.

6.2.2 ATM standardization process

ATM took shape through the full international consultative procedure, although the standard specifications do not resolve every possible issue. This process took about ten years (1985-1995). Researchers in the 1980s were looking at the best method to achieve high-speed packetized data transmission. It began with ISDN standards activities. ISDN is fully digital, but it is characteristic of its generation. Its physical layer (level 1) is based on copper wire, and its bandwidth is limited to hundreds of thousands of bits per second. ISDN has been popular in Europe and Japan. The European Union has stimulated a program for universal or near-universal ISDN availability before the end of this decade. Adoption in the United States has been more spotty, although some regional telecommunication companies are turning to ISDN as a relatively low-cost means to increase bandwidth to subscribers.

ISDN allocates the available bandwidth into three channels, using time division multiplexing (also known as *synchronous transfer mode*) technology: two 64 Kbps channels for data transport and one 16 Kbps channel for signaling (housekeeping messages). The signaling channel is used to establish a connection; the data transport channels then support transfer of the information. ISDN is strictly a circuit (Nx64, 1=N=30). ISDN was not designed to support broadband applications.

A decision was made in the mid-1980s to seek a new standard, which could be based, to some extent, on ISDN principles and support optical fiber; because of the medium used, the supported speeds are much higher. The newcomer became known as B-ISDN (for Broadband ISDN). Although both are digital, ATM technology differs distinctively from ISDN. ISDN is a *synchronous transfer mode technology* (i.e., ISDN is a circuit-switched technology without any statistical or multiplexing and statistical gains); ATM, on the other hand, is an *asynchronous transfer packet technology* with statistical multiplexing and gain. Because the user and the carrier *gamble* on statistical multiplexing, sophisticated traffic management capabilities are required.

Fundamental ATM concepts arose from research conducted in the mid-1980s. This work established that data units of fixed length were easier to switch at very high speeds than frames (such as Ethernet, Token Ring, and FDDI) which could vary in length. This philosophy was influenced by research done at IBM on a device known as the Packetized Automatic Routing Integrated System (Paris) switch, which serviced both variable- and fixed-length packets. The Paris project assumed, however, that all traffic would consist of data. It was a rationing card system, in which senders could transmit only if they had received a token, which was awarded on the basis of an average data flow. (IBM later abandoned Paris and embraced ATM.)

Several entities have published relevant standards, specifications, and requirements, including the International Telecommunications Union (ITU), Telecommunications (ITU-T), and ANSI (ATIS). The ATM Forum (ATMF), the Frame Relay Forum, and Bellcore have published relevant documentation. Carriers require the support of the ATMF User-to-Network Interface Specification version 3.1 (with version 4.0 to follow in 1996), Broadband Intercarrier Interface (B-ICI) version 2, and (for internal connectivity only) Private Network Node Interface (P-NNI) version 1.0 or higher. UNI 3.1 supports the ITU-T Q.2931, which is important. LAN Emulation version 1.0 or 2.0 is also generally required.

The B-ISDN debates that led to standardization were carried on between three factions. One faction, which could be identified as the X.25 faction, was concerned above all with more efficient, versatile, and less-costly data transmission. The data faction was relatively unconcerned about time of arrival and could tolerate some delay while packets were reassembled. They were very concerned about protecting the integrity of the information. The public telephone providers, representing the third faction, emphasized time-sensitive information such as voice, in which consistent sequence is crucial. Limited information loss was acceptable. There were, however, common grounds. For packetized video, packets must come through at a regular rate in order to avoid transmission "jitter." This is easier to accomplish with relatively compact packets. The factions solved this problem by deciding on a fixed-length packet, called a *cell*, which could be transmitted in an orderly, high-speed fashion over a switched network. This solution would provide the cost advantages of data networks combined with the predictability of voice networks. But how long should those cells be? The data faction advocated a 64-byte specification, while the voice faction demanded 32 bytes. An agreement was finally reached: each cell would contain a 48-byte payload, accompanied by five additional bytes to identify the cell, and carry other protocol control information.

For high-bandwidth networks based on fiber, ATM is frequently employed with a layer 1 standard, known as SONET.* SONET defines a series of bandwidth levels for transmission over fiber networks. SONET rates consist of multiples of a base of 51.840 Mbps. Current ATM technology supports bandwidths at the OC-1 (Optical Carrier-1) level (51.840 Mbps), OC-3 level (155.520 Mbps), and OC-12 level (622.08 Mbps). SONET levels now targeted by system developers would deliver 1244.160 Mbps (OC-

* Outside the United States, the SONET concept is described in the context of the Synchronous Digital Hierarchy (SDH); effectively, this hierarchy uses building blocks of 155.520 Mbps rather than building blocks of 51.840 Mbps. However, they are basically consistent for appropriate values of the aggregate bandwidth.

24) and 2488.320 Mbps (OC-48). SONET standards are now nearly ready for OC-192 (about 10 Gbps) speeds. (*Note*: For user access, current ATM technology only supports the 155.520 Mbps; the 622.080 Mbps speed is only currently supported at the trunk level.)

OC-24 and OC-48 would represent the gigabyte network concept mentioned so often in public discussion. Applied with ATM technology, these bandwidth levels should be achievable later in this decade.

6.2.3 ATM as a practical technology

ATM's architecture creates possibilities that have been beyond the reach of earlier technologies. As discussed, accredited standards bodies have developed the basic set of protocols. As is always the case, implementers' workshops are then needed to complete the implementation details. In the case of ATM, the ATM Forum has focused on issues of interoperability. It appears that ATM will resolve these issues more quickly than was the case for its predecessors. This is primarily due to two factors: focus and commitment. The ATM Forum consists of a plethora of vendor, user, government, and academic representatives, whose commitment to the success of ATM is, in some people's view, unparalleled. FDDI, for example, has encountered some early difficulties in linking workstations from different manufacturers. Token Ring also had problems of this kind; Ethernet, however, encountered fewer interoperability obstacles because of the inherent simplicity of its architecture. This was due to the lack of an organization focusing efforts on implementation and deployment issues.

Because it can utilize fiber media to its full potential (at least in the long haul) ATM supports high speed. *Note*: In the campus environment, ATM uses multimode fiber to derive 100 Mbps (using FDDI-like encoding) or single-mode fiber to derive 155 Mbps (using FCS-like encoding), thereby not utilizing the fiber to its full potential. Also, it uses UTP 5 twisted pair to support speeds up to 155 Mbps—hence, ATM provides high speed when, and only when, it uses the underlying medium to the full potential. These larger transmission capacities, used in conjunction with appropriate switching technology, make it easier to cope with bursty traffic. Ethernet, with its 10 Mbps ceiling, can also handle bursty traffic but is hindered by the contention issue and the easy-to-reach ceiling. As discussed, there are transmission (now 622 Mbps) and switching (now 20 Gbps) ceilings in ATM. But these are, at this time, more difficult to overwhelm. Token Ring and FDDI reduce contention compared to Ethernet but have much less capability to handle bursty transmissions. ATM's ability for "dynamic" allocation of bandwidth makes it easier to offer different classes of service to support different application classes. The transparency of the upper layers to ATM, when appropriate adaptation is provided (in pertinent equipment), makes it possible for ATM to be ubiquitous, employed in multiple types of LAN and WAN topologies and thus in every segment of the end-to-end connection.

ATM can be economically deployed in specific corporate environments, and its cost effectiveness will increase over time. Error checking is performed only on the cell header, and not on the payload. This means that error-free reception of payload is not guaranteed at the ATM layer (it will be guaranteed at the TCP/IP layer). This is based on today's high-quality, low-noise transmission media. Fiber media, used with the SONET Physical Layer protocol, makes available high capacity to the connected users. However, there is still a need to support statistical multiplexing and bandwidth overbooking, in order to obtain transmission efficiencies, particularly in the long haul. It follows that in ATM, sophisticated congestion control is needed in order to "guarantee" a very small probability of cell loss even under significant traffic levels. These characteristics add up to a technology that can permit users and applications developers to explore possibilities that have not been feasible until now.

6.3 Overview of key ATM features

This section outlines some of the principal features of ATM; later chapters will describe some of these in greater detail. The features discussed here are as follows:

- The structure of its 53-byte cells, or labeled information containers
- The physical layer, ATM layer, and ATM adaptation layer, which organize appropriate Service Data Units (SDUs)/PDUs for transmission. Special attention is given to the adaptation layer, which governs the treatment of cells that require different qualities of service to accommodate the special requirements of voice, video, and data traffic. The service layer sits on top of the AAL and uses specific AALs (e.g., AAL 1, AAL 5, etc.) to provide the appropriate services to the legacy protocols (e.g., IP) residing at the network layer.
- LAN Emulation, in support of legacy LANs.

ATM is a set of standards, defined originally by the ITU-T. These standards establish basic specifications for ATM protocols and interfaces. The ITU-T standards for ATM specify the cell size, structure, and the UNI. Note that there are two kinds of UNIs: one for access to public networks and one, called *private UNI*, for access to a customer-owned ATM network (specifically to a hub, router, or switch). For the public UNI, the physical layer is defined for data rates of 1.544 Mbps, 45 Mbps, and 155 Mbps (SONET OC-3). For the private UNI, a number of physical layers for different media (UTP, STP, single-mode fiber, and multimode fiber) are defined.

ATM can be described as a packet transfer mode based on asynchronous time division multiplexing and a protocol engine which uses small fixed-length data units known as cells. ATM provides a connection-oriented service (although in theory it can also be used to support connectionless services such as Switched Multimegabit Data Service [SMDS]). Note that LANs such as Ethernet, FDDI, and Token Ring support a connectionless service. Each ATM connection is assigned its own set of transmission resources; however, these resources have to be taken out of a shared pool, which is generally smaller than the maximum needed to support the entire population—this is the reason for the much-talked about traffic management problem in ATM. ATM, nevertheless, makes it possible to share bandwidth through multiplexing (multiple messages transmitted over the same physical circuit). Multiple virtual channels can be supported on the access link, and the aggregate bandwidth of these channels can be overbooked (ATM relies on statistical multiplexing to carry the load). Within the network, expensive resources are "rationed," and bandwidth must be allocated dynamically. ATM is thus able to maximize resource (bandwidth) utilization.

A virtual circuit can be either switched (temporary) or permanent. A connection is established through preprovisioning with the carrier or private devices (thereby establishing Permanent Virtual Channels [PVCs]), or through signaling mechanisms (thereby establishing Switched Virtual Channels [SVCs]). Connections supported by these channels (PVCs or SVCs) enable one computer or other system on the network to communicate with another.

Network resources such as inbound speed, outbound speed, quality of service, multipoint capabilities, and so forth, are requested as a connection is established. A connection is established if the network is able to meet the request; if not, the request is rejected. Once the virtual circuit is defined, the call connection control assigns an interface-specific Virtual Channel Identifier (VCI) and Virtual Path Identifier (VPI) to identify the connection. These labels have only interface-specific meaning. Two different sets of VPIs/VCIs are assigned to the two end points of the connection. Inside the network, as many sets of VPIs/VCIs as needed (along the path) are used by the network, invisible to the end users. As long as the connection remains active, the assigned VCI and VPI represent valid pointers into routing tables in the network; the tables (accessed via the VPI/VCI) are used to accomplish cell routing through the network.

What is the distinction between PVCs and SVCs? Communication in ATM occurs over a concatenation of virtual data links called *Virtual Channels*; this concatenation is called a Virtual Channel Connection (VCC). VCCs can be permanently established by an external provisioning process, entailing a service order (with desired traffic contract information) and manual switch configuration. Such a VCC is known as a Permanent Virtual Connection. When the control plane mechanisms are implemented in both the user equipment and in the switch (specifically ITU-T Q.2931 and/or ATMF 3.1), the user will be able to establish connections automatically on an as-needed basis. This type of connection is called a Switched Virtual Connection (SVC).

What is the difference between VC switching and VP switching? ATM supports two kinds of channels: Virtual Channels (VCs) and Virtual Paths (VPs). VCs are communication channels of specified service capabilities between two (intermediary) ATM peers. Virtual Channel Connections (VCCs) are a concatenation of VCs to support end-system-to-end-system communication. VPs are groups ("bundles") of VCs. Virtual Path Connections (VPCs) are a concatenation of VPs to support end-system-to-end-system communication. In VC switching, each VC is switched and routed independently and separately. VP switching allows a group of VCs to be switched and routed as a single entity. This concept only applies to ATM and not to the other services available over the ATM platform.

6.3.1 The ATM cell

The ATM cell has a 48-byte payload, accompanied by a five-byte header, which is divided into fields. Headers are of two types: the UNI and the Network Interface (NNI). See table 6.1.

Table 6.1 ATM Cell Structure

8	7	6	5	4	3	2	1
Generic Flow Control				Virtual Path Identifier			
Virtual Path Identifier (continued)				Virtual Path Identifier (continued)			
Virtual Channel Identifier				Virtual Channel Identifier			
Virtual Channel Identifier (continued)							
Virtual Channel Identifier (continued)				Payload Type Identifier			Cell-loss Priority
Header Error Control							
Payload (48 bytes)							

Fields within the UNI cell are as follows:

- The first field, consisting of four bits, provides for Generic Flow Control (GFC). It is not currently used and is intended to support a local bus ("extension") function to connect multiple broadband terminal equipment to the same UNI as equal peers (note that multiple users can be connected to the UNI today by using a multiplexing—not peer—function). This is equivalent to the SAPI function in ISDN.

- A 24-bit routing pointing field is subdivided into an 8-bit VPI subfield and a 16-bit subfield for VCI. It indirectly identifies the specific route laid out for traffic over a specific connection, by providing a pointing function into switch tables that contain the actual route.
- Three bits are allocated to the Payload Type Identifier (PTI), which identifies whether each cell is a user cell or a control cell used for network management.
- A single-bit Cell-Loss Priority (CLP) marker is used to distinguish two levels of cell-loss priority. Zero identifies a higher-priority cell that should receive preferred loss treatment if cells are discarded due to network congestion. One indicates lower-priority cells, whose loss is less critical.
- Header Error Control (HEC) is an eight-bit cyclic redundancy code (CRC) computed over the ATM cell header. The HEC is capable of detecting all single-bit errors and certain multiple-bit errors. It can be used to correct all single-bit errors, but this is not mandatory. This mechanism is employed by a receiving device to infer that the cell is in error and should simply be discarded. It is also used for cell-boundary recovery at the physical layer.
- The remaining 48 bytes are devoted to payload (in case of interworking, some of the payload bytes have to be set aside for network-used AAL protocol control information).
- The NNI cell structure has one difference. The four-bit GFC field is dropped, and the VPI field is expanded from eight bits to 12.

6.3.2 Addressing

Addressing is a fundamental need in any network. The ITU-T ATM protocols call for a hierarchical ISDN telephone numbering scheme, specified in ITU-T E.164, to be used in ATM. The standard permits the ATM address to be divided into an address and a subaddress. The ATM Forum recommends that the address describe the point of attachment to the public network (if connected to the public network) and that the subaddress identify a particular end station within a private network [5]. Note that VPI/VCI are just labels, not E.164 addresses; they are table pointers for the relaying of cells on to their destination, based on switch routing tables.

The ATM Forum specification permits two address formats to be used as an ATM address. One is the E.164 format, and the other is a 20-byte address modeled after the address format of an OSI Network Service Access Point (NSAP), as seen in table 6.2.

Table 6.2 The ATM Forum's Address Format for Private ATM Networks

1 octet	2 octets	1 octet	3 octets	2 octets	4 octets		6 octets	1 octet
AFI	DCC/IDC	DFI	AA	Reserved	RD	Area	ESI	SEL
Authority and Format Identifier	Data Country Code/ International Code Designator	Domain-Specific Part Format Identifier	Administration Authority		Routing Domain		End-System Identifier	(Unused)
					(Selected by Organization)			

Note: The end-system identifier contains a valid IEEE 802 MAC address. An alternative format allows the eight bytes after the AFI field to contain an E.164 address. This option permits both public and private subaddresses to be combined into a single ATM address.

6.3.3 The physical and ATM layers

An extension of the conventional OSI seven-layer stack can be used to describe the structure of the ATM protocol: a reference model specific to ATM depicts its structure more clearly. This reference model distinguishes three basic layers. Beginning from the bottom, these are the physical layer, the ATM layer, and the AAL. Each is divided further into sublayers, as seen in table 6.3.
The physical layer includes two sublayers.

1 As with any other data link layer protocol, ATM is not defined in terms of a specific type of physical carrying medium, but it is necessary to define appropriate physical layer protocols for cell transmission. The Physical Medium (PM) sublayer interfaces with the physical medium and provides transmission and reception of bits over the physical facility. It also provides the physical medium with proper bit timing and line coding. There will be different manifestations of this layer, based on the specifics of the underlying medium (e.g., DS1 link, DS3 link, SONET, UTP, etc.).

Table 6.3 ATM Reference Model

Convergence	CS	AAL
Segmentation and Reassembly	SAR	
Generic Flow Control (if/when implemented) Cell VPI/VCI Translation Cell Multiplex and Demultiplex		
Cell Rate Decoupling HEC Header Sequence Generation/Verification Cell Delineation Transmission Frame Adaptation Transmission Frame Generation/Recovery	TC	PHY
Bit Timing Physical Medium	PM	

2 The Transmission Convergence (TC) sublayer receives a bit stream from the PM sublayer and passes it in cell form to the ATM layer. Its functions include cell rate decoupling, cell delineation, generation and verification of the HEC sequence, transmission frame adaptation, and the generation/recovery of transmission frames.

The ATM layer, in the middle of the ATM stack, is responsible for one of ATM's most "trivial" functions: to encapsulate downward-coming data into cells from a number of sources and multiplex the cell stream; conversely, it is responsible for deencapsulating upward-coming cells and demultiplexing the resulting stream out to a number of sources.

The ATM layer controls multiplexing (the transmission of cells belonging to different connections over a single cell stream) and demultiplexing (distinguishing cells of various connections as they are pulled off the flow of cells). ATM, as a data link layer protocol, is medium-independent: it is capable of performing these functions on a wide variety of physical media. In addition, the ATM layer acts as an intermediary between the layer above it and the physical layer below. It generates cell headers, attaches them to the data delivered to it by the adaptation layer, and then delivers the properly tagged cells to the physical layer. Conversely, it strips headers from cells containing data arriving on the physical layer before hoisting the data to the application layer.

VCs and VPs are identified by their VCI/VPI tags. Do not confuse VCs/VPs with connections. VCs are channels; connections are instances of end-to-end communication. Connections are identified by call reference and connection identifiers included in the setup message used in signaling. See reference [2] for a more extensive description. The ATM layer assures that cells are arranged in the proper sequence, but it does not identify and retransmit damaged cells. If this is to be done, it must be accomplished by translated VCI/VPI information.

Each ATM switch has its own routing table to identify each connection. In transit between switches, VPI/VCI identifiers (routing table pointers) will be different. Switches translate identifiers as they transfer cells onward to other switches.

Finally, the ATM layer performs management functions. If the Payload Type Identifier (PTI) identifies a cell as a control packet, the ATM layer responds by carrying out the appropriate functions.

6.3.4 Class of service: the adaptation layer

The ATM adaptation layer allows various network layer protocols to utilize the service of the ATM layer. As discussed earlier, the ATM layer only supports the lower portion of the data link layer. Hence, in order for the network layer to use ATM, a *filler sublayer* is required. This is analogous to IP use over a LAN: the Media Access Control layer only supports the lower portion of the data link layer; consequently, the logical Link control layer is sandwiched in between.

Fundamentally, the AAL keeps the network layer "happy" by enabling it to use ATM transparently. The basic function of the AAL is to segment the downward-coming data (network layer PDU) into cells and to reassemble upward-coming data into a PDU acceptable to the network layer. It is critical to understand that AALs are end-to-end functions (end system-to-end system). A network providing pure ATM will not be aware of, cognizant of, or act upon AAL information (only in case of service interworking in the network does there have to be network interpretation of the AAL information).

In one classical view of the ATM protocol model, a "service" layer resides above the AAL, in the end system. Hence, by further elaboration, one can say that in a coincidental manner, the AAL differentiates the treatment of different categories of cells in the end system and permits responses to user-to-user quality-of-service issues. A number of AALs have been defined to meet different user-to-user quality-of-service requirements. Again, a network providing pure ATM will not be aware of, cognizant of, or act upon AAL information. Therefore, the AAL-supported service differentiation is among end-system peers and is not the mechanism used by the ATM network to support network quality of service. We use the term *user-to-user quality of service* to describe the end system-to-end-system peer-to-peer connection service differentiation (this connection being viewed as "external" to the ATM network).

An end-system TV monitor, for example, needs a continuous bit stream from a remote codec in order to paint a picture; it may have been decided that an ATM network is to be used to transport the bits. Because of the codec/monitor requirements, the bits have to be enveloped in such a manner that clock information is carried end to end so that jitter is less than some specified value. To accomplish this, the bits are enveloped using AAL 1. From the ATM network's point of view, this is totally immaterial: the ATM network receives cells and carries them to the other end; the network delivers cells. The network does not render any different type of QoS to these cells, based solely on the fact that the cells had AAL 1 information in them; the network was not even aware of the content. Naturally, it would be desirable if the network provided reasonable QoS to this stream, based on some kind of knowledge or arrangement.

The different qualities of service obtained via an ATM network are based on user-to-network negotiation, not by the content of the cell. In PVC this negotiation is via a service order. Here, the user would tell the network (with paper) that the user wanted to get reasonable service for a certain stream carrying codec video. The network provider would make arrangements to terminate this stream on a switch line card where, for example, a lot of buffers were allocated. The network provider then tells the user (with paper) to employ a certain VPI/VCI combination (say 22/33) for this specific stream. Here is what happens: The user sends cells over the physical interface terminated at the card's port. Certain cells arriving on the interface have VPI/VCI = 44/66; these get some kind of QoS treatment. Then some cells arrive on the interface with VPI/VCI = 22/33; these cells get the agreed-upon QoS by receiving specific treatment from the switch. In SVC, a similar mechanism is in place, except that instead of communicating the information using paper, the call-setup message is used (with automatic call negotiation).

In any event, the QoS in the ATM network is not based on the fact that the cells carried a certain AAL. It is the other way around. The user needs a certain end-system-to-end-system QoS. The user then needs to do two things: select its (network-invisible) AAL and separately inform the carrier of the type of QoS needed.

AALs utilize a (small) portion of the 48-byte payload field of the ATM cell by inserting additional control bits. In all AALs, the ATM header retains its usual configuration and functions. Notice that the data coming down the protocol stack are first treated by the AAL by adding its own header (protocol control information). This AAL PDU must naturally fit inside the ATM PDU. Hence, the AAL header must fit inside the payload of the lower layer, here ATM. To say that quality-of-service definitions are obtained at the cost of reductions in payload is not exactly correct. AAL provides an appropriate segmentation and reassembly function—QoS is supported by the network switch. As discussed, AAL classes support peer-to-peer connection differentiation.

In some instances, users determine that the ATM layer service is sufficient for their requirements, so the AAL protocol remains empty. This occurs, for example, if the network layer protocol can ride directly on ATM (this is unlikely for legacy protocols), or if the two end systems do not need additional coordination. In the majority of cases, however, this AAL layer is crucial to the end-system protocol stack, because it enables ATM to accommodate the requirements of voice, image/video, and data traffic, while providing different classes of service to meet the distinctive requirements of each type of traffic.

Two sublayers make up the AAL: the Segmentation and Reassembly (SAR) sublayer and the Convergence Sublayer (CS). The SAR sublayer segments higher-layer information into a size suitable for cell payloads through a virtual connection. It also reassembles the contents of cells in a virtual connection into data units that can be delivered to higher layers. Functions such as message identification and time/clock recovery are performed by the CS sublayer. See table 6.4.

Table 6.4 ATM Class of Service

	Class A	Class B	Class C	Class D
Application	Voice Clear Channel	Packet Video	Data	
Timing (Source-Destination)	Needed		Not Needed	
Mode	Connection-Oriented			Connectionless
Bit Rate	Constant	Variable		

The ITU-T specifications apply three broad criteria to distinguish four classes of ATM service, tagged as A, B, C, and D; these end-to-end (network-external) criteria are as follows:

1. Time relation between source and destination
2. Bit rate
3. Connection mode

In order to express these criteria in practical form, four AALs have been developed: AAL 1, AAL 2, AAL 3/4, and AAL 5, focused upon data transmission. The four end-to-end (network-external) classes of service are as follows:

1. Class A (e.g., clear-channel voice and fixed bit-rate video, such as movies or high-resolution teleconferencing): A time relation exists between source and destination. The bit rate is constant, and the network layer-level service is connection-oriented.
2. Class B: As in Class A, there is a time relation between source and destination; network layer-level service is connection-oriented, but the bit rate can be varied. Examples include audio and video with variable bit rates (e.g., unbuffered video codecs with motion compensation).
3. Class C: The network layer-level service is connection-oriented, but there is no time relation between source and destination and the bit rate is variable. This can, for example, meet the requirements of connection-oriented data transfer and signaling.

4 Class D: Intended for applications such as connectionless data transport (at the network layer); none of the three parameters applies. Service is connectionless, there is no time relation between source and destination, and the bit rate is variable.

These classes are general descriptions of types of user traffic. They do not set specific parameters or establish values. Equipment from multiple vendors based on different parameters may thus make it difficult to establish connections. AALs are end to end and generally external to the ATM network; considerations on AAL relate to considerations of end user equipment. AALs are considered by the network only when there is service interworking. Examples include frame relay-to-ATM interworking in the network, legacy LAN-to-ATM interworking (specifically, LANE) in the network, and private line-to-ATM interworking in the network. In the first case, frames come in and cells go out.

There is another case where AALs are used in the network, but this is totally transparent to the user. This situation (called by some "network interworking") is when the network supports a "carriage function" over ATM. Examples include frame relay carriage over an ATM network, Ethernet carriage over an ATM network (e.g., Ethernet bridging), and private line carriage over an ATM network. The frame relay user gives a frame relay frame to the (ATM-based) network; the network takes the frame and segments it into a stream of cells utilizing AAL 5 protocols. The stream is carried across the network and in proximity of the destination; the cells are reassembled into a frame relay frame using AAL 5. The destination is handed a frame. This type of service is called "frame relay carriage over ATM" or "frame relay-to-ATM network interworking."

So far, three AAL protocols have been defined to support the three classes of service in the end-system: AAL1, AAL 3/4, and AAL 5. Computers, routers, and other devices must employ the same AAL in order to communicate with one another on an ATM network.

- AAL 1 meets the performance requirements of Service Class A. It is intended for voice, video, and other constant bit-rate traffic, and its performance, to the upper layers of the end-system stack, is similar to today's digital private lines. Four bits in the payload are allocated to Sequence Number (SN) and Sequence Number Protection (SNP) functions.
- AAL 2 aims at Class B requirements.
- AAL 3/4 is intended for connectionless data services (e.g., for support of switched multimegabyte data service). Four bytes are devoted to control functions, including a multiplexing identifier as well as Segment Type (ST) and SNP indicators.

- AAL 5 is also intended for data communication, including services such as frame relay, LANE, and MPOA. The ATM Forum and IETF recommend that AAL 5 also be used to encapsulate IP packets in the user's end system. AAL 5 is specifically designed to offer a service for data communication with lower overhead and better error detection. It was developed because computer vendors realized that AAL 3/4 was not suited to their needs. In addition to the header, AAL 3/4 takes an additional four bytes for control information from the payload field, reducing its capacity by 8.4 percent. Vendors also maintained that the error-detection method of AAL 3/4 does not adequately cope with issues of lost or corrupted cells. See table 6.5.

Table 6.5 AAL Type 5 CS-PDU

Information Payload	PAD	Control	Length	CRC-32
0–64K	0–47	1 Byte	2 Bytes	4 Bytes

With AAL 5, the CS sublayer creates a CS Protocol Data Unit (CS-PDU) when it receives a packet from the higher application layer. The first field is the CS information payload field, containing user data. The PAD field assures that the CS-PDU is 48 bytes aligned. A one-byte control field remains undefined, reserved for further use. The two-byte length field indicates the length of information payload, and the CRC field is used to detect errors.

When the CS sublayer passes the CS-PDU to the SAR sublayer, it is divided into many SAR Protocol Data Units (SAR-PDUs). The SAR sublayer then passes SAR-PDUs to the ATM layer, which carries out transmission of the cell.

When passing on the final SAR-PDU within the CS-PDU, SAR indicates the end of the CS-PDU transfer by setting to one the payload type identifier (PTI) in the header. By using the CS length field and the Cyclic Loss Redundancy Code (CRC) in the header's HEC, the AAL can detect the loss or corruption of cells.

6.3.5 LAN Emulation

Lane Emulation (LANE), is an interworking capability. It allows Ethernet/Token Ring stations to communicate directly with ATM stations (and vice versa) as if they were using the same protocol. The interworking equipment supports the conversion between the two protocols.

Traditional LANs use the 48-bit MAC address. The MAC address is globally unique. This non-hierarchical LAN address, assigned by the manufacturer, identifies a network interface in the end station. The use of a MAC address is practical in a single LAN segment or in a small internet. However, large bridged networks become difficult to manage and experience large amounts of broadcast traffic for the purpose of attempting to locate end stations. The address space of a large network is preferably hierarchical. This makes it easier to locate a particular point on the network; such an address, however, restricts the mobility of the network users. The E.164 address used in public ATM is hierarchical.

To emulate a LAN, the ATM network must support addressing using the MAC address scheme: each ATM MAC entity must be assigned a 48-bit MAC address, from the same address space, to facilitate its identification. As noted, an ATM network, whether public or private, uses a hierarchical address. The address resolution operation in LANE binds the end station MAC address to the physical address of the ATM port to which the end station is currently connected. When an end station is attached to an ATM switch port, a registration protocol exchanges the MAC address between the ATM network and the end station.

The LAN Emulation service consists of several pieces of software and hardware operating on one or more platforms. Prior to explaining the operation some definitions are necessary.

- LAN Emulation Client (LEC): The LEC is software that resides at the edge device. The edge device is where the emulated service is rendered in terms of the conversion between protocols.
- LAN Emulation Server (LES): The LES provides initialization and configuration functions, address registration, and address resolution. Since ATM and legacy LANs use very different addressing schemes, a way to map the two is important, particularly with a view to subnetworks, where the addressing capabilities of ATM may be lacking.
- Broadcast and Unknown Server (BUS): The BUS provides the mechanism to send broadcasts and multicasts to all devices within the emulated LAN.

In a traditional LAN, all frames (unicast, multicast, and broadcast) are broadcast to all stations on the shared physical medium; each station selects the frames it wants to receive. A LAN segment can be emulated by connecting a set of stations on the ATM network via an ATM multicast virtual connection. The multicast virtual connection emulates the broadcast physical medium of the LAN. This connection becomes the broadcast channel of the ATM LAN segment. With this capability, any station can broadcast to all others on the ATM LAN segment by transmitting on the shared ATM multicast virtual connection. This issue is treated in more detail in chapter 11.

6.4 Narrowband ATM access

The ATM protocol is most efficient when operating at high speeds. As stated before, these cells are 53 bytes long, with 48 bytes of payload and 5 bytes of header information. This is almost a 10 percent overhead for the header information, just at the ATM layer (there are other inefficiencies at the AAL and physical layer). The ATM-level inefficiency is not considered excessive when operating at high speeds, given the ability to mix voice, video, and data.

Currently, low-speed access links are the norm for most WAN environments. ATM's overhead becomes burdensome at these speeds, typically 56 Kbps to T1 (1.544 Mbps). The ATM Forum addressed this requirement by developing a new UNI aimed at increasing the efficiency for low-speed links. This new UNI is called the Frame UNI (F-UNI) and operates on frames that can have payloads of up to 4,096 bytes. Frame relay, Data Exchange Interface (DXI), and F-UNI are all frame-based standard.

- Frame relay defines HDLC as part of its specification and adds a header to support Data Link Control Identifier (DLCI) addressing.
- DXI/UNI is an evolution of the ATM DXI (e.g., see reference [1]), which defines router interfacing to an ATM CSU (through HSSI), although the DXI/UNI also defines V.35 and N x 64,000 (via a regular CSU). For N x 64,000, frames are carried all the way to the switch at the central office. The AAL 5 convergence sublayer and VPI/VCI addressing scheme are used. The ATM DXI/UNI allows a customer to access a network supporting ATM technology based on HDLC frames. The purpose of this interface is to provide HDLC access to ATM at low speeds.
- F-UNI is separate from DXI/UNI. It can be thought of as a superset to DXI/UNI. The difference between the two is that F-UNI also extends Q.2931 signaling for SVCs. It also carries AAL 5 convergence sublayer PDUs. Hence, F-UNI defines an alternate HDLC-based protocol for user access to ATM.

By using ATM DXI, one can support legacy DCEs by encapsulating frames and giving them to the central office interworking unit, which then prepares them for ATM transport. This approach is good for PVCs, but not for SVCs. For SVCs, one should use F-UNI. Here the same encapsulation is used for the user plane and for the control plane; it entails a dual-stack interworking unit at the switch/central office.

The F-UNI supports full ATM signaling, enabling frame-oriented devices such as bridges and routers to set up switched circuits and negotiate class of service with the network. The F-UNI header contains flags to indicate standard ATM features, such as cell-loss priority, congestion notification, and the presence of operations and maintenance traffic. The F-UNI specification details the appropriate mapping function to go between frame headers and cell headers. It also supports adaptation layers 3, 4, and 5. For management, F-UNI supports the Interim Local Management Interface (ILMI).

6.5 Issues regarding WAN services

How does a carrier provision an ATM service? ATM (and ATM-based) services entail five levels of provisioning. First, appropriate physical lines must be deployed to the sites in question (this can entail more than two user locations and the physical bandwidth to each location can be different). Second, the central office switch must be configured (port, slot, switch capacity, trunk capacity, etc.). Third, the concentrator must be configured (port, slot, switch capacity, trunk capacity, etc.). Fourth, the QoS parameters (SCR, PCR, BT, service type, etc.) must be configured into the network management station. Fifth, the virtual connectivity must be specified (the PVCs/SVCs, workgroups, etc.).

When is a remote concentrator used? LAN bridging, LAN Emulation, and FDDI bridging require a remote concentrator that supports the AAL functions. Frame relay can be done over an access line (but currently a non-channelized-only service is supported). Native ATM UNI service does not generally require a remote concentrator, unless there are enough customers at one location to make it more effective to use a concentrator/ multiplexer in common space than to use a multitude of transmission facilities.

How is bandwidth allocated in an ATM network? For ATM services, the customer can specify (the equivalent of) the Sustainable Cell Rate (SCR), the Peak Cell Rate (PCR), and the Burst Tolerance (BT). The switch allocates various resources (e.g., trunk capacity and buffers) on a statistical basis. This involves the use of traffic shapers, traffic policers, and tagging for discard. The carrier has the obligation to deploy enough resources in the network to guarantee the type of quality of service and the kind of service required (e.g., constant bit rate, variable bit rate, available bit rate, unspecified bit rate, etc.). Non-ATM services are allocated in a similar manner, but the user has no direct control on the (ATM) traffic parameters.

What is meant by ABR, VBR, UBR, and CBR? For native ATM, a number of services are available in support of different user requirements. CBR provides a constant bit rate in support of a service that provides the equivalent of a private line at T1 or T3 rates. VBR is a variable bit-rate service in support of data applications such as frame relay and LAN interconnection. ABR provides available bit-rate service at a discount basis; here the amount of bandwidth in the network is not guaranteed and is managed via a feedback mechanism. UBR is an unspecified bit rate with only "best-effort" characteristics.

Physically, how are different services supported? ATM services are supported over a fiber DS3/OC-3 link; the user adds equipment supporting the ATM peer. Frame Relay services are supported over a T1 line; the user adds equipment supporting the Frame Relay peer. LAN-based services are supported by installing the concentrator at customer sites. The concentrator is connected to the network over an appropriate ATM trunk; the user connects to the concentrator over an SAS/DAS FDDI connection.

What user equipment is needed to access a carrier's cell relay service and/or LAN bridging service? For ATM cell relay, the user can use an ATM-ready hub, ATM-ready router, or workgroup switch. For a legacy service, the user delivers an Ethernet or FDDI connection to the carrier's site-located concentrator.

What is the relevance of P-NNI? Does it matter that our switch does not communicate over the P-NNI to another switch? Carriers have not, heretofore, supported access by the user to the trunk side of their switches (e.g., voice switches, packet switches, frame relay switches, and SMDS switches). They will not likely allow access to the trunk side of an ATM switch. Carriers may only offer a UNI service, not a P-NNI service. Hence, it does not matter that a switch may not interwork with another switch over the NNI—this is not relevant to the service a carrier provides. The only relevance of P-NNI is in the event the carrier decides to use P-NNI as an NNI to connect equipment *within* the carrier's network (but not outside of it).

Does ATM support multimedia; and if so, how? Video and multimedia information can be carried either over an ATM UNI, an FDDI interface, or an Ethernet interface. The ATM switch supports a high grade of service, which makes it a good platform for these media.

Does ATM support multipoint services, and how? ATM supports multipoint connections. The ATMF UNI 3.1 specifies the use of signaling messages to establish such connections. Many ATM switches do not currently support ATM UNI multiconnections (they do support a kind of internal multipoint service for LAN support).

6.6 References

1. D. Minoli and M. Vitella, *Cell Relay Service and ATM for Corporate Environments* (New York: McGraw-Hill, 1994).
2. D. Minoli and G. Dobrowski, *Principles of Signaling for Cell Relay and Frame Relay* (Norwood, MA: Artech House, 1995).
3. D. Minoli and T. Golway, *Designing and Managing ATM Networks* (Greenwich, CT: Manning Publications/Prentice Hall, 1997).
4. D. Minoli, *Telecommunications Technologies Handbook* (Norwood, MA: Artech House, 1991).
5. P. Newman, *ATM Local Area Network, IEEE Communications* (March 1994).

CHAPTER 7

Quality of Service in ATM Networks

Thus far we have discussed the basics of ATM and described applications that make use of the technology. The preceding chapters were designed to provide the foundation necessary to understand the applications and protocols driving data communication. Throughout the remainder of this book we will focus on how ATM technology can be used to build next-generation, multiprotocol, high-speed networks that support different levels of user-selectable quality of service. In mapping out the future of data communication we will walk through several different technologies and, in the concluding chapters, describe how they are integrated for a multiprotocol over ATM network.

In order to build networks that support next-generation applications, there are several building blocks, qualities of service, and resource reservation protocols that are required. The remainder of this book will discuss these protocols in a phased approach. This chapter is the starting point of that explanation and will begin with a discussion of what quality of service, QoS, is and how it can be measured. QoS provisions allow an application to specify resource requirements and to reserve resources to ensure correct operation or better operation than would be achievable without the feature.

The feature of providing QoS is critical for next-generation networks, because it is the technology that allows effective sharing of the network resources among dissimilar applications. ATM, due to its ability to support and segregate different multimedia applications, is an enabling technology for integration of services, and for this reason it continues to gain interest and popularity in both the LAN and WAN environment. In a basic migration strategy, most existing applications, such as data communication, can be carried on ATM by interworking the technologies and adapting data formats. Next-generation applications, such as high-quality video conferencing or voice communication, can directly utilize QoS on an ATM network by negotiating the desired network behavior through the ATM Application Protocol Interface (API).

QoS mechanisms [2] are primarily targeted at applications operating on real-time data, such as audio and video transmission. Another use for differentiating QoS levels that is becoming more popular is to segregate a higher-grade service for corporate users interconnecting their intranet or extranet sites across the Internet. When the ISP's routers interconnecting corporate intranet sites are used to carry both mainstream Internet traffic along with extranet flows, the ISPs can achieve significant returns on their capital investment. The key enabler in this model is the ability to divide the traffic within the network into multiple categories and then discard the lower-priority commodity Internet traffic during times of congestion. The ability to have a corporation's traffic carried across the WAN without loss, but also achieving some economy of scale, is a very interesting service, which comes at a premium price and has become important to the business cases of most ISPs. (See figure 7.1.)

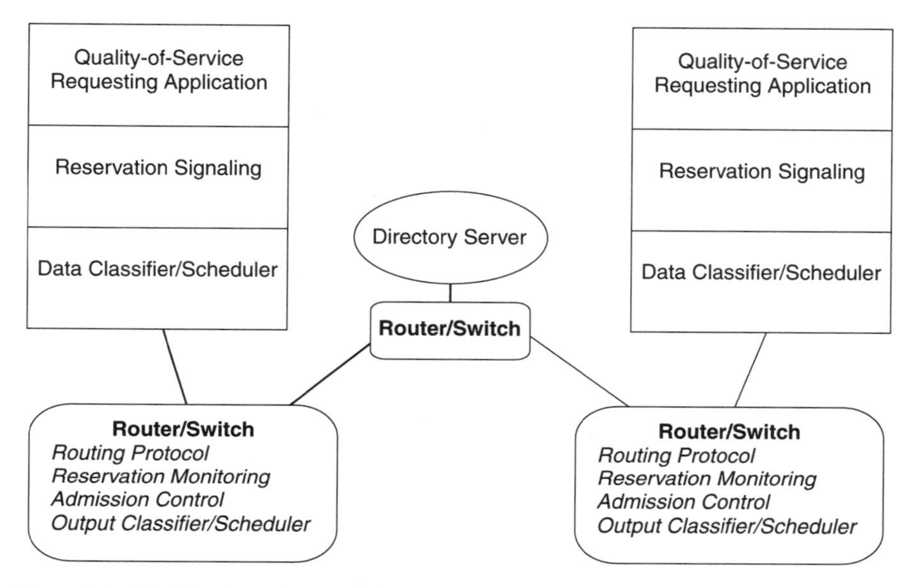

Figure 7.1 High-level, end-to-end QoS

In this chapter we will examine the issues and definitions of QoS at an advanced level. We will discuss the parameters used to define QoS from both the ATM perspective and the legacy network's perspective, describe the ATM signaling messages used to negotiate QoS, and explain ATM's service categories used to carry QoS-sensitive traffic. This information forms a foundation for the discussion in subsequent chapters of new protocols that can use ATM and next-generation IP networks as a QoS-rich fabric.

7.1 ATM QoS and traffic management

ATM has been designed as a technology that allows resource sharing between various applications on the network, and sharing is one of the primary economic selling points for network design when selecting the technology. Consequently, ATM has been positioned as one of the most flexible and widely deployed technologies in computer communication. ATM attempts to achieve higher utilization in two ways:

- By statistically multiplexing multiple similar traffic types onto one connection using either a best-effort or fair-sharing algorithm
- Achieving higher utilization by segregating traffic of different priorities and guaranteeing delivery of the higher priorities (e.g., protecting a video VCC from a data

VCC)—thus, allowing congested network connections to reach maximum capacity while still passing mission-critical data.

The goal of statistical multiplexing is to run several different data streams through one physical link and get better total utilization than if time division multiplexing were used. When the network is carrying traditional best-effort Internet-type traffic, not all hosts will be using the network at all times, so statistically there is a good chance that their data will be passed successfully. If, on the other hand, all of the sources were to transmit simultaneously, serious congestion would occur without resources being reserved.

When the total available bandwidth on a link is overallocated, then the link utilization can be very high. However, high utilization comes at the price of poor-quality service, as perceived by the hosts that were concurrently transmitting data and, thus, congesting the link. If these hosts are not well equipped to deal with congestion, they will continue to overload the link and effectively destroy each other's data. Traditionally, hosts using TCP have used its end-to-end congestion-avoidance mechanisms to provide equitable bandwidth sharing. Bursty data applications that use ATM networks today, without congestion feedback, are especially prone to congesting the network quickly.

Ironically, it is often likely that the reason for compounding the problem is higher-layer protocols, such as TCP, which detect data loss and request retransmissions, thus further congesting the network. The problem is linked to the cell loss spread out over many individual packets, which makes a small congestion event affect potentially many packets. If, for example, the cell loss was one in every hundred, but the average packet length required 100 cells, then the effective throughput would be zero.

In order to better cope with the problem of congestion, a major focus of the ATM Forum was to devise a scheme that dynamically changes the rate at which computers transmit data into the network, thus attempting to equally share the bandwidth among all the hosts. This effort in the ATM Forum's traffic management group resulted in the Available Bit Rate (ABR) specification and a related effort known as Early Packet Discard (EPD). Both are addressed later in this chapter.

In ATM, QoS comes from intelligent traffic management, it constitutes one of the functions performed at the ATM layer, and controls the rate that traffic enters the network by *shaping,* or pacing, the flow. The parameters used by the ATM layer to pace traffic are those defined for the virtual circuit at call setup time or when the PVC was created by a network administrator. The main objectives of ATM layer traffic management are as follows:

- Provide predictable behavior to achieve performance objectives
- Minimize congestion

- Maximize the efficient use of network resources
- Operate independently of data being carried or the AAL used.

The operation of an ATM layer protocol, such as ABR, is to support current applications and, hopefully, most future bursty data sources. In addition, the ATM layer is made up of procedures and protocols that are designed to not overtax the host computer or routers. This point has not yet been proven on very large-scale networks but, with the ever-increasing performance of ATM interfaces, the goals should be achievable.

When a host or router creates a virtual circuit, it does so via signaling messages which specify the source, destination, and other parameters for the QoS desired. The QoS parameters are derived from a collection of traffic models that describe source flows and schemes for transmitting data into the network. These are compliant with the data traffic model's definition (i.e., ATM Forum traffic classes specified in the circuit creation). In the following sections, the service categories will be discussed from the viewpoint of their effect on traffic management.

7.1.1 ATM's QoS parameters and call setup/routing

The QoS parameters are values that help to quantify the traffic being transmitted into the ATM network across a virtual circuit. The parameters are specified at virtual circuit creation time and are used to establish the traffic contract. The traffic contract is an agreement between the data source and the network in which the source can know with some certainty that data transmitted within the bounds of the QoS parameters will be delivered to the destination.

At virtual circuit creation time, the ATM network uses the traffic parameters to determine if there are sufficient resources to accept the call. This process is known as Call Admission Control (CAC) and is tightly coupled with ATM's interswitch routing protocols, the most important of which is the Private Network Node Interface (PNNI). PNNI is responsible for determining if the required resources exist across the entire ATM network.

CAC is a process executed by ATM switches as they receive new call setup messages, either from the source computer or from another upstream ATM switch. The intent of CAC is to determine if, on a switch-by-switch basis, adequate resources exist in the switch to support the requested quality of service. When a switch that has received a call setup message determines that it does indeed possess adequate resources to accept the call, it must then reroute the message to the next ATM switch along the best path to the ultimate signaled destination.

The determination of which path should be used to route the call setup message is one of the key functions of the PNNI interswitch routing protocol and is based on two criteria:

1. The shortest path between source and destination is considered (i.e., the conventional routing problem).
2. The switch must calculate the likelihood of successfully routing the call among all of the possible paths between it and the destination. The most likely path is the one that will have sufficient resources for subsequent switches to accept the call.

This second criterion is addressed with a protocol known as Generic Call Admission Control (GCAC) and is an ATM Forum-supplied algorithm switches can run to determine the likelihood of their multivendor neighbor switch accepting a call setup message. The combination of the shortest-path routing coupled with QoS aware/capable paths is one of the problems addressed in the PNNI specification.

QoS aware routing is a critical technical problem to solve, because QoS requirements that are specified by applications and communicated to the network via the ATM signaling protocol are guaranteed at virtual circuit creation time through route selection and resource reservation. The best path is chosen by the interswitch virtual circuit routing algorithm that is QoS aware—that is, it understands that different paths leading to the same destination may provide very different QoS characteristics. Therefore, virtual circuits requiring high QoS must be very selective in the path chosen to cross the network.

The problem regarding path selection is only half the battle in getting QoS aware applications to run on an ATM network. The next problem is that of mapping network layer protocols to ATM's connectionless fabric. Internetworking between ATM and IP is being solved, in part, by providing IP best-effort services over ATM networks using Classical IP (RFC 1577) and the ATM Forum's LAN Emulation. Although the problem of IP routing is supported via these approaches, as will be discussed in chapter 10, both of these solutions fail to make the ATM real-time service facilities available to IP applications.

An integrated model that provides QoS capabilities to applications and utilizes an ATM network will, in all likelihood, be implemented by combining a QoS negotiation protocol, such as that covered in chapter 9, with the ATM Forum's Private Network Node Interface (PNNI) protocol. PNNI is an ATM routing protocol designed to support QoS on a global ATM Internet. The protocol is divided into two sections:

1. PNNI routing, which sustains hierarchical, link state, and source-based decisions with QoS support
2. PNNI signaling, which uses the ATM Forum UNI specification as a foundation, allowing ATM switches to exchange reachability information.

The key to the operation of the protocol and its ability to support QoS is centered on PNNI's support for hierarchical network views. At a high level of abstraction, the protocol operates by subdividing the total set of interconnected ATM switches into subgroups. The subdivision is usually done on departmental boundaries (i.e., engineering, marketing, etc.) first, and then on administrative or corporate boundaries (i.e., XYZ Corporation). Within each group, a top-level switch, or leader, is selected to act as the spokesperson for QoS negotiation on behalf of the complete set of switches in that layer of the subdivision. The set is called the *peer group,* and the top switch is referred to as the *peer group leader.* Each peer group can then exchange state information, detailing if a link has failed and/or circuit utilization, to establish knowledge of network load and reaffirm the consequences of establishing new connections with strict QoS requirements.

The peer group leader uses the concept of hierarchical views when it examines the state of its members and passes that information on to its adjoining peer group leaders. The outcome of this aggregation and dissemination of information is that individual ATM switches can establish connections across the ATM network with a high degree of certainty that desired QoS requirements will be met. The other major benefit of this process is scalability. Because the peer groups distill information as they work their way up the hierarchical view, the total work required to maintain the data is manageable.

The following describes the step in using PNNI for virtual circuit creation. When connections are established in a PNNI environment, a computer or router attached to an ATM switch will issue a call setup message, with its associated QoS values, and pass the message to its local ATM switch. The switch using the global knowledge gained by being a member of the PNNI peer groups will examine the call setup message. The switch will determine which path has the highest probability of meeting or exceeding the desired QoS.

This is based on its understanding of the current state of the network, the shortest path, and the resources being requested. Afterwards, the switch forms the path through the network as it passes the call setup message to the next switch. Because the first ATM switch that the host or router encounters attempts to calculate the entire path for the connection, the PNNI protocol is source-based routing. As the call setup message traverses the network, it is passed to peer group leaders that check their current state and forward it to the appropriate next hop. In the case where the state of the network has changed, so that a switch along the path is unable to meet the QoS requirements, the call is cranked back one hop and the path calculation is repeated. (See figure 7.2).

Once the circuit is created, the network can use the traffic description parameters to police the stream of switch input cells to determine if the user is transmitting at the agreed-upon rate. When the switch accepts a call and agrees to establish a virtual circuit, it also agrees to transport all compliant cells belonging to the virtual circuit. Compliance

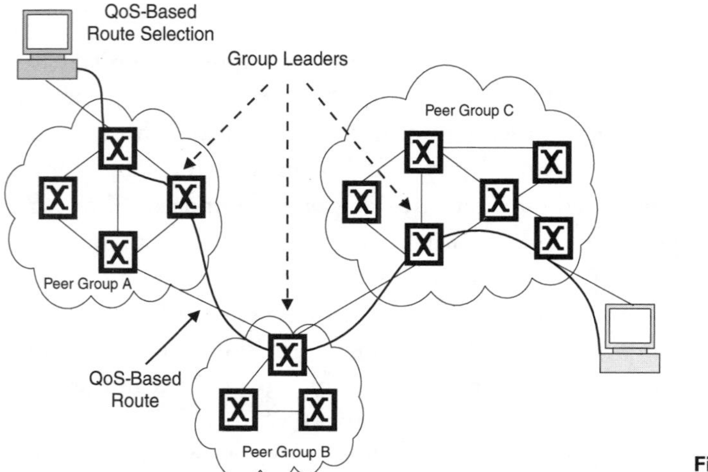

Figure 7.2 PNNI

checking is termed Usage Parameter Control (UPC) and involves validating several aspects of the virtual circuit. It is the application of the UPC that ensures QoS for all users of the ATM network.

ATM switches and IP routers that support QoS by policing their input traffic implement what is called the *Leaky Bucket* algorithm. The Leaky Bucket provides a method of describing the rate of traffic when data are transmitted into the network in a frame or cell format. Each Leaky Bucket has an input rate and a limit parameter. The actual policing is done via vendor-specific code, and the Leaky Bucket is just a concept to help visualize the algorithm of policing.

Call admission control uses the traffic parameters to make a best estimate if the switch, or switches, can carry data between clients at the desired bandwidth. ATM QoS is specified to the software implementing the call admission control as the performance that can be realized by the end devices via the negotiation of the following traffic description parameters.*

- Peak Cell Rate (PCR)—the maximum number of cells that can be transmitted over a unit time
- Cell Delay Variation Tolerance (CDVT)—the amount of *clumping* that can be tolerated before a series of cells exceeding the PCR is deemed as violating the traffic contract

* The traffic parameters are defined by the ATM Forum in the User Network Interface (UNI) specification.

- Sustainable Cell Rate (SCR)—a rate of cells that is near or slightly higher than the average over a unit time
- Maximum Burst Size (MBS)—the of cells on a virtual circuit that can burst at the PCR
- Minimum Cell Rate (MCR)—the minimum rate at which a host using the available bit rate service category will always be able to transmit data. The default value is zero. When the minimum rate is set to zero, the network does not always promise that bandwidth will be available.

A successful call setup, and subsequent virtual circuit creation, involves selecting a set of values from the above parameters that are acceptable to the CAC software running on the switches. If the desired resources exist, then the switch will respond to the end devices that call is established. The successful set of parameters is called the "traffic contract" because it specifies that the network will make some guarantees about performance if the network resources utilized are less than or equal to those specified in the call setup request.

After the call has been established, the Usage Parameter Control (UPC) software on the switch monitors the resources being used to determine if the traffic contract is be obeyed. Any UPC algorithm that is policing the maximum offered load of the traffic contact uses the combination of the PCR and CDVT. The policing done on the values of the PCR and MBS is done in the first Leaky Bucket. This bucket is allowed to accept logical cells until it overflows, in which case the overload feeds into subsequent measurement buckets.

In some cases, picking the values for the traffic parameters can be very difficult, because most applications, such as legacy LAN traffic, are not traditionally specified by these metrics. And in many cases network managers either specify no policing of LAN traffic or set the values of PCR/SCR only because their Internet provider is selling limited bandwidth. In that case, the ISP can utilize ingress policing on the ATM ports serving customers to check that only the allocated bandwidth is used.

For a virtual circuit requiring high QoS, or at least a very consistent stream of cells, the peak rate could be nearly identical to the bandwidth allocated. If the virtual circuit is carrying bursty data, the peak rate requested is typically somewhat lower than the actual maximum, because the call is more likely to be accepted and the transmitting host's ATM layer should be able to shape the egress cell stream.

The SCR and MBS are the parameters used in constructing the second Leaky Bucket measurement device. The SCR is always chosen by the end system when signaling for the creation of the virtual circuit and is, by definition, lower than the PCR. MCR is used only with the available bit rate traffic category. When considering SCR values, the network manager should consider values that will provide consistently good performance to applications over long periods of time. That way the data rate will be able to burst to the PCR for short periods but always be able to depend on SCR performance.

The traffic parameters just discussed are used by the network to determine which resources are required for the virtual circuit. In addition to those listed, the following QoS parameters are signaled in a call setup message. (See table 7.1.)

Table 7.1 Application of QoS Performance Parameters

QoS Performance Parameter	QoS Assessment Criteria
Cell Error Ratio	Accuracy
Severely Erred Cell Block Ratio	Accuracy
Cell Misinsertion Ratio	Accuracy
Cell Loss Ratio	Dependability
Cell Transfer Delay	Speed
Cell Delay Variation	Speed

- Cell Loss Ratio (CLR) = lost cells to transmitted cells and specifies how much traffic has been lost compared to the total amount transmitted. The loss can be attributed to any cell-corrupting event such as congestion or line encoding errors.
- Cell Transfer Delay (CTD) is the measure of the time required for the cell to cross certain points in the ATM network. The primary concern to end users is the time required, for the last bit of the cell to leave the transmitter until the first bit arrives at the receiver. CTD can be effected by processing in the ATM switches or, more likely, buffering in the switch.
- Cell Delay Variation (CDV) is the measure of how the latency varies from cell to cell as the cells cross the network. Queuing causes variation and CDV is also caused by variation in switching speeds that cells encounter as they are transmitted. CDV is of concern in ATM networks, because as the temporal pattern of cells is modified in the network so is their traffic profile. If the modification is too large, then the traffic has the potential of exceeding the bounds of the traffic profile, and some cells may be dropped, through no fault of the transmitter.

The signaled QoS parameters just discussed above are used in the following manner when requesting service.

- CLR applies to CBR, real-time VBR, and non-real-time VBR.

- For ABR, a value of CLR may be associated with the service, but it is not signaled.
- CTD is carried in the call setup messages for CBR and real time VBR services.
- CTD is carried in the call setup messages for CBR and real time VBR services.

There are several QoS parameters provided by the network that are not signaled but are nonetheless very important when used to differentiate service. Typically, these values are specified by a service provider for different products or by switch vendors for different switches. This means that the parameters are a point that can be used for differentiation of service—for example, switch vendors may claim that their products are superior because of lower values on average cell transfer delay. Also, some carriers differentiate their services over the competitors by advertising lower values of cell loss or transfer delay—hence, the ultimate quality of service delivered to the end user is superior.

Nonsignaled parameters are generally understood to be quantitative values which are published quality descriptions applicable to all services of the network. It is also possible that an ATM service provider or vendor may specify different values of these parameters for different types of services available. In this case the general application of various degrees of services can be differentiated by a few clearly different steps in the QoS; realized. Then the steps are marketed as several supported service-level guarantees. It may even be the case that the customer is buying Ethernet connectivity, but different grades of ATM service are used to provide the various service-level guarantees.

Switch/service differential parameters are as follows:

- Cell Error Ratio (CER) = erred cells/(transferred cells + erred cells)
- Cell Misinsertion Ratio (CMR) = misinserted cells/time interval
- Severely Erred Cell Block Ratio (SECBR).

7.1.2 Effectiveness of traffic parameters

Traffic parameters can be divided into three categories, which illustrate their effect on traffic or the ATM QoS realized. The QoS assessment criteria differentiate the service from the following standpoints:

1. Speed, which translates into latency and affects applications such as video conferencing that require a short delay for images to be transmitted from one participant to another.
2. Accuracy, which reflects different quanta of errors and would negatively affect and/or be important to application, such as file transfers or extranet communication.
3. Dependability, which is an umbrella term that reflects the overall accuracy.

From the preceding discussion of traffic parameters, call admission control, and usage parameter control, there is a subtle hint that the most favorable input data stream to an ATM switch or network is one with a smooth, consistent data flow. This is true of all networks but may be particularly true for ATM when the data virtual circuits are specified in terms of PCR, SCR, and MBS.

Because smooth traffic is a desired trait, it is important that the client ATM software (i.e., traffic shaping at the ATM layer) on the traffic source correctly generates and transmits data traffic that is traffic contract compliant and, even better, shaped to be smooth. To make good use of ATM, a host will be required to internally buffer the output stream to comply with its traffic contact. In some cases, such as transmitting data from a bursty video source, this may be difficult, but smoothing to a certain degree should be possible. There may also be examples where the call setup will fail with high-traffic parameter values, and in order to successfully create the virtual circuit, the hosts need to scale back their bandwidth requests.

In traffic shaping, constant bit-rate applications shape their traffic using the PCR traffic descriptor. In this case, the offered load will be scheduled at one cell every 1/PCR unit of time. A variable bit-rate source shapes its bursts to the PCR rate. Additionally, it must shape its traffic to ensure that no more than the MBS will be transmitted at the PCR. A closed loop, feedback controlled, rate-varying source requires dynamically shaped output at all times. The rate-varying mechanism complies with the traffic contact by adjusting its output to match the value supported by the network at that moment.

7.2 ATM service categories

The preceding discussion has covered call admission and what traffic parameters are used in describing virtual circuit performance. In this section, we address the ATM Forum's service categories, which are used to help define five different QoS classes, these are supported by the current ATM Forum User Network Interface (UNI) specification.

The service architecture, with five categories, is defined in the ATM Forum's Traffic Management (TM) 4.0 specification. One of the main concepts presented there is that provisioning virtual circuits based on the large set of traffic parameters can become very complex and the number of permutations is huge. In addition, the ATM Forum has realized that end-user applications, such as video conferencing or WWW browsing, tend to fall into certain application categories; these can be used to make general statements about QoS needs and requirements on the network.

If the above simplification used when provisioning ATM services, then the set of variables is reduced and network management can be greatly simplified. Therefore, by placing traffic into different categories, the user benefits because it restricts the amount of information specified when the virtual circuit is created. In the TM 4.0 model, when a virtual circuit needs to be created for a service, such as a constant bit-rate source, then only a subset of the total possible traffic and QoS parameters will be transmitted in the call setup message.

In the TM 4.0 specification, the ATM Forum has defined the behavior of each category and identified the applicable signaling parameters. The service categories are also useful tools for ATM switch designers, because they allow designers and manufacturers to consider each traffic type differently and potentially optimize their products based on the traffic mix they feel will be most common. This allows policing, priority, and scheduling algorithms to be developed, with special performance enhancement for the category of interest.

The major improvements of the TM 4.0 specification over previous UNI specifications were the addition of the ABR text and the clarification of the service categories. Prior to the TM 4.0 work, the only text addressing traffic management was contained in the original ATM Forum UNI 3.0 specification. The UNI 3.0 document is primarily concerned with congestion-avoidance mechanisms, while the TM 4.0 document contains language that addresses how to provide differentiated QoS more efficiently.

It is worth noting that prior to the TM 4.0 work, traffic management at call setup time was governed by the user's combination of seven parameters: PCR, CDVT, SCR, MBS, CLR, CTD, and CDV. This led to complex and confusing virtual circuit creation messages which were disliked by both vendors and users.

It is also important to keep in mind that the work in the ATM Forum on QoS categories is an ongoing process and may be extended and refined with time. Extensions are being proposed for the best-effort category to support an enhancement that will have minimum bandwidth guarantees. In order for ATM to provide high QoS for all anticipated services in the TM 4.0 specification, each model would need to be supported.

The following list defines the ATM Forum's service categories:

- Constant Bit Rate (CBR) has a deterministic bandwidth requirement which makes it easy to provision/engineer. Sources that can make use of CBR generate cells at a fairly consistent rate, fully occupying the allocated bandwidth most of the time. The source rate corresponds to a known peak emission rate measured in Cells Per Second (CPS). Conforming cells are typically guaranteed high priority, because they are the product of latency-sensitive traffic, such as video, voice, and circuit emulation. Loss of cells or large delay will have a substantially negative impact on these applications.
- Variable Bit Rate Real Time (VBR-rt) is bursty data with delay sensitivity.
- Variable Bit Rate Non-Real Time (VBR-nrt) is bursty data without delay sensitivity.

- Available Bit Rate (ABR) is the ATM Forum-cell-based adaptive flow control.
- Unspecified Bit Rate (UBR) is a best effort category.

In order to understand how these service categories are used, the following text gives examples of how the categories are recommended in the TM 4.0 specification. The CBR category would be applied to very consistent traffic sources—for example, DS1 circuit emulation. The VBR-rt category would be used for applications such as video conferencing or video distribution, which run at a high priority. The remainder of the service categories would be used for data applications, and the particular category selected would most likely depend on the personal preferences of the network manager. In the following section, we will elaborate on this point from the perspective of applying the ABR and UBR service categories to data transmission for legacy networks.

7.2.1 ABR and UBR for packet traffic

As described previously, ABR is a service category that tries to realize better all-around network performance for all ABR clients on the ATM network by modifying each client's output to suit the amount of traffic the network can tolerate at that particular time. ABR support for clients allows them to modify their output and also to increase their offered load when there is no congestion.

With adaptive flow control, it is believed that an ATM network is capable of supporting high-quality services, such as CBR, which has fixed bandwidth requirements and variable bit-rate services, such as bursty data, which are content with the leftover bandwidth. Historically, telecommunication applications have called for fixed bandwidth allocations, since voice communication generated a constant data stream.

Conversely, data communication has a long history of using protocols, such as TCP that can adjust the amount of data transmitted to network congestion. A slow start will sense congestion by detecting packet loss and then reduce the size of bursts of traffic transmitted. The ABR specification is an example of the ATM Forum developing a networking technology that is designed to be used simultaneously by telecommunication carriers and enterprise data networks.

ABR is designed to implement an adaptive mechanism which modifies its output speed depending on feedback from the network. By using an adaptive protocol, ABR's problem of determining in advance what load will be offered and then statically allocating bandwidth for the length of the communication, regardless if it is used or not, is eliminated. Applications make use of all available bandwidth on the network but reduce the offered load when congestion occurs. With ABR, hosts sharing the network understand and correctly respond to the feedback messages, so they will realize a fairer distribution of resources, and a higher quality of service.

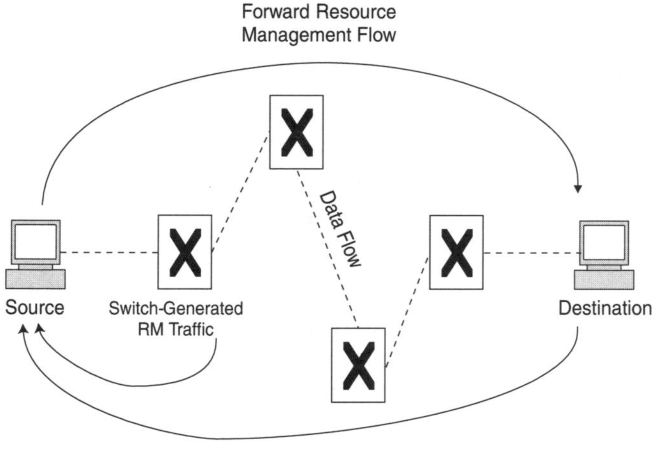

Figure 7.3 ABR RM flows

An ABR source dynamically varies its transmission rate by reacting to special cells known as Resource Management (RM) cells. RM cells are consistently generated by the traffic sources at regular intervals, regardless of the existence of congestion. During normal noncongested operation, the RM cells are passed between the source and destination unmodified, which denotes that the network's resources are not overloaded. The RM cells are then returned to the source as if they made a "U" turn on the ATM network. However, when congested, the RM cells are modified—that is, their bit pattern is changed to signify congestion either in the network or at the destination device.

When a source receives an RM cell indicating congestion somewhere between the source and destination ATM clients, the source then reduces its transmission rate by one-half of the current bit per second average. Because the protocol attempts to maximize the utilization of network links, the source next attempts to slowly increase its output speed until it receives another RM cell indicating congestion. This transmission rate increase/decrease interaction between the ABR source and RM cells results in appropriate adjustments in the transmission rate to avoid congestion. (See figure 7.3.)

In order to make certain that the source is obeying the RM request for bandwidth use reduction, it is necessary to police the transmission at the ATM switches. In addition, the ATM switches will need to monitor their buffer utilization and each port will need to create RM cells where applicable.

There are some areas of concern with ABR that will remain until more practical experience using the protocol has been gained, such as the response time of the protocol, multicast support, and its interaction with existing congestion-avoidance transport protocols such as TCP/IP.

The concern over the response time has to do with the ability of the ABR hosts to quickly respond to RM cells when increasing or decreasing their offered load and determining at what rate the change occurs. The speed at which a host increases its load to the network will determine how fast it gets bandwidth on demand. If the host ramps up slowly, then the QoS will be poor. If the host ramps up too quickly, the switch buffers may overflow before back-pressure RM cells can be generated.

Another concern with ABR is its inability to work when used with multicast VCCs. Because ABR utilizes a closed-loop feedback mechanism with the loop formed between the source and destination, it is not possible to have different transmission rates on a one-to-many multicast circuit. If multiple receivers share the single multicast VCC, then their feedback loops will merge and their RM cells will not be differentiated by the source. If the source cannot determine where the RM cell comes from, it will not be able to scale back its transmission rate—consequently, the system will fail. Until future versions of the protocol are developed to address this problem, however, the only option currently available is to create unique virtual circuits for each source destination pair.

Finally, there is a potential problem if ABR is used on an ATM network with other virtual circuits that are not using ABR. In this situation, when the network begins to experience congestion, it notifies the ABR circuits to reduce their offered load. However, the ATM switches have no way of communicating congestion avoidance to the non-ABR virtual circuits. Therefore, the ABR users would follow the rules and slow down, while the non-ABR users would utilize the available network bandwidth. Clearly, this will not inspire widespread deployment of ABR, if several users are still utilizing the widely deployed UBR service category.

The UBR service category is the ATM Forum's attempt to model a traffic type similar to LANs or the Internet. The UBR class is "best effort" because it offers no quality-of-service guarantees. Data are transmitted by the hosts with the understanding that the network will do its best to carry the information to the destination, but, due to congestion or other detrimental situations, data may be lost. From the standpoint of policing, the UBR service can be thought of as similar to VBR, with the PCR and SCR equal. However, if the two coexist on virtual circuits crossing the same ATM switching fabric, UBR is typically given a lower priority.

UBR is an attractive service class, because it is simple to support on ATM switches and conforms to existing philosophies of the Internet. The switches do not need to perform any resource reservation or precall setup bandwidth allocation algorithms. This dramatically reduces the complexity of transmitting traffic, maintaining state/accounting while a call is in progress, and it also reduces the complexity of designing and building

ATM devices. UBR is also attractive from the host's perspective, because it does not require prior knowledge of traffic—a critical shortfall is the forced selection of PCR/SCR/MBR. UBR places no ATM layer traffic-management egress-shaping requirements on ATM devices and relies on higher-layer protocols, such as TCP, to react to congestion.

A final component which is tightly coupled to UBR and its similarities to Internet-like performance, is the concept of full-frame discard, sometimes called Early Packet Discard (EPD), in a congested ATM network. Frame discard is a relatively simple technique, which allows ATM switches to discard entire AAL5 frames. It can be used with UBR to effect a service quality very similar to the current Internet services (i.e., routers transmitting IP packets).

When a packet is segmented with AAL5, the last cell of the segmented PDU has a bit set in the payload type indicator, to indicate that the packet has been completely processed and transmitted. If the ATM switch is experiencing congestion and needs to discard cells, it can now use this AAL5 marker to drop cells that belong to only one packet. EPD has the advantage that, in times of bursty congestion, only one virtual circuit at a time is losing traffic. Unlike ABR, EPD also has the advantage of functioning equally as well with multicast data and has been shown to perform very well in reducing congestion with both multicast and unicast traffic.

A potential drawback to EPD is that because it uses the PTI bit, which is set only when using AAL5, EPD is very tightly coupled to that ATM adaptation layer. If a host application needed to use a different AAL, say for AAL1 encoded voice, then the hardware in the ATM with EPD switch would not have the ability to determine the end-of-packet marker and would fail to improve performance at times of congestion. EPD also relies on higher OSI layers for flow control, but not all transport layer protocols can provide this service. Therefore, protocols that do not have the ability to pace themselves when operating on congested networks may not benefit from EPD.

7.3 The Internet's use of quality of service

In addition to the complexity of measuring the use of switched virtual circuits, there is an equally complex problem of negotiating quality of service with a common language used by all applications. Over the last several years, the Internet Engineering Task Force (IETF) has been examining support for QoS over a packet-switched network and has attempted to address several of the QoS-related problems.

This work has been split between two IETF working groups: the Resource Reservation Protocol (RSVP) and the Integrated Services (int-serv) group. When building a system that supports QoS, the RSVP specification is the mechanism that performs QoS requests, which is similar to ATM signaling. The int-serv specifications concern documenting the capabilities available to QoS-aware applications, which is similar to ATM traffic management.

For the most part, the two standards bodies, the ATM Forum and IETF, have developed similar and compatible specifications. Each technology supports the basic mechanisms for QoS, such as traffic descriptors, policing policy, admission control policy, resource allocation, and scheduling algorithms. However, some of the terminology used by the IETF is different from that of the ATM Forum, and the differences should be understood before describing the service categories.

The IETF has defined service categories that fall into one of the following divisions. These categories are important, because they are the most likely services ISPs will support on their networks, regardless of whether or not the underlying fabric is ATM technology.

- Guaranteed service allows the user to request a maximum delay bound for an end-to-end path across a packet-switched network. Service is guaranteed to be within that delay bound, but no minimum is specified. The ATM service most similar to this is CBR.
- Controlled load provides a small number of service levels, each differentiated by delay behavior. It supports three relative levels, with no particular numerical values of delay associated with them.
- Best-effort service represents the service that can be achieved over the Internet without any QoS modifications. This category is the default.

The critical challenge is how to couple the popular IETF packet-based technology with the emerging ATM Forum cell-based technology.

7.4 Measuring QoS

Measuring QoS is becoming increasingly controversial and may need more technological and policy development for years to come. In many cases during the standardization process in the ATM Forum, network providers have been reluctant to include strong wording when referring to guaranteed QoS, because it may allow customers to dispute service quality. This is a very real concern to the carrier, because of the large range of possibilities in ATM signaling messages and the somewhat unknown behavior of ATM switching equipment. Nevertheless, the complex signaling messages have been standardized, and the ability of carriers to supply service that meets that requested of the ATM virtual circuit will be an area of market differentiation.

When an ATM network is used to transport data for an Internet application, the various ATM QoS metrics, such as cell transfer delay, cell delay variation, cell-loss ratio, and so forth will have varying degrees of impact upon the performance perceived by the user. A key area of interest in the research community is the analysis of how the ATM QoS parameters can be used to describe or measure the actual performance realized by the end user when the end user is generating frame-based traffic such as TCP/IP. The following are possible areas that can be measured and their relationship to ATM.

- One of the first performance measurements is the analysis of throughput. Throughput is the measure of the rate at which the offered load can be transmitted across the network and can be further subdivided into lossless and peak throughput. Lossless specifies the maximum data rate with no loss, and peak is the maximum data rate during loss. A final clarification is sometimes called *goodput* (the amount of good data that is actually transmitted) and is used to define the actual amount of usable data that reaches the end user over a unit of time. Probably the most significant negative factor for IP traffic is a high level of the CLR. If the CLR is excessive, the erred packets at the IP layer will typically be an order of magnitude larger than the error count at the ATM cell layer, because of the difference in packet size versus cell size. (See figure 7.4.)

- Frame delay, as with cell transfer delay, is measured when a frame crosses the ATM network. The ATM Forum has defined the Message-In Message-Out (MIMO) term to describe this measurement. MIMO is the amount of time between the last bit of the frame entering the network and the first bit of the frame leaving the network. The value of frame delay is that it measures the ability of an ATM switch or network to pass data efficiently. Most of the groundwork on frame latency was completed in the IETF and is documented in RFC 1242 [1]. When applied to the

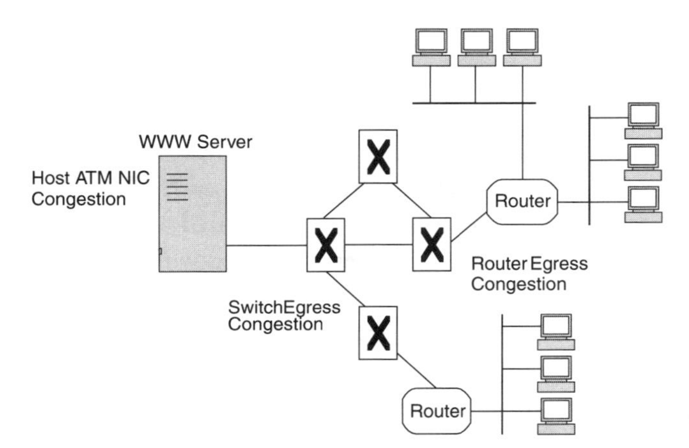

Figure 7.4 Different points of congestion versus throughput

ATM protocol suite, frame delay is the latency beginning when the frame is passed to the ATM adaptation layer prior to segmentation and ending when the first bit of the frame is received at the AAL peer. When measured as the time required to pass frames between hosts, the service category used and the amount of load on the ATM network will affect the frame delay.

- Frame-loss ratio is the counterpart to cell-loss ratio and is defined as the number of frames lost in the network due to congestion or other errors.
- Frame loss ratio = (input frame count – output frame count)/(input frame count)
- Maximum frame burst size is the counterpart to the cell layer maximum burst size and is the measure of the number of frames that can be sent at the peak rate without loss.
- Virtual circuit setup latency is the measure of the time required to set up a switched (i.e., on demand) virtual circuit between the source and destination. This measure is important for short-lived communication because if the number of frames is small and could be transmitted over a connectionless network in less time than required to set up a virtual circuit, the user will perceive a lower QoS. Because this measurement takes into account the delays incurred as each switch processes the call setup message and routes the call, this value will be dependent upon the size and construction of the ATM network. This value is useful in comparing the performance realized in connection-oriented versus connectionless (i.e., pure IP router) networks. (See figure 7.5.)

7.5 Additional complexity of cell-based accounting

From the previous discussion, it should be clear that an ATM virtual circuit can be complex, because of the large number of permutations the traffic-provisioning parameters can take. However, the capability to signal a traffic parameter value for each of the QoS options does not necessarily mean that the feature will be supported or that the requested value will be supported. There are two key problems involved with generated complex call setup messages.

First, the range of possibilities is so large that service providers may be forced to restrict the set of requested values. By reducing the quantization of the range-requested values, the provider can gain better control on the problem. If a network offers two

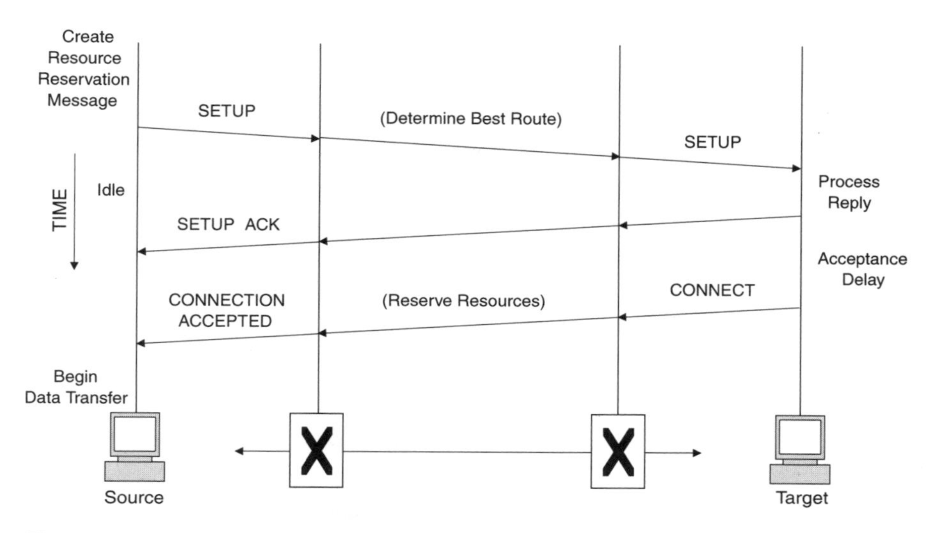

Figure 7.5 Virtual circuit setup latency

grades of VBR service with CLR of 10^{-4} and 10^{-7}, and two ATM-attached workstations request CLR values of 10^{-5} and 10^{-6}, then both would need to be given the better service.

A second major problem to be considered when signaling complex call setup messages is the call accounting procedures. ATM call setup messages present the unique problems of both maintaining large amounts of information upon creation and the generation of additional data to track what resources were actually used during the call.

Until large-scale switched virtual circuit deployment is more common, the magnitude of this problem is unknown. However, it is widely believed that current network management and carrier billing systems will experience difficulty in supporting the amount of accounting data an ATM network will generate, due to the large number of virtual circuits used in data communication. Thus, new models of data collection that aggregate at intermediary hosts, along with billing based on generalized parameters, will be used.

In order to make use of switched virtual circuits in an environment where accounting data needs to be collected for billing and/or network monitoring, it is likely that some new model of data collection will be used. Some vendors have proposed models where only a small number of "buckets" are used to place virtual circuit statistics. In this model, the stored counters are placed in the bucket that is numerically closest to the virtual circuit. This type of aggregation can dramatically reduce the amount of data that each switch must maintain per customer when providing a use-based service.

7.6 Summary

This chapter has provided an introduction to the concepts and terminology behind QoS, virtual circuit creation, and QoS measurement parameters. Because of the complexity of ATM and QoS in general, technology advances from this point onward may be relatively slow due to the lack of experience in measurement techniques and provision policies. Currently, little practical experience has been gained about what can be specified in an achievable QoS request, because there is uncertainty about what can be delivered by the networking hardware. This problem is compounded by difficulties in current ATM-based ISPs' abilities to offer different service-level agreements.

Nevertheless, it is important to realize that a great deal of groundwork has been accomplished to facilitate deployment of QoS-aware applications and, as will be shown in subsequent chapters, to influence the network infrastructure to be QoS-aware. The ATM Forum has generated a signaling and traffic specification in the UNI 4.0 release, which describes QoS and standardizes machines that could potentially be used to deliver the service. The IETF has also been very active in this area and has developed documents describing QoS achievables on frame-based networks, along with their version of a signaling protocol: RSVP.

In the following chapters we will describe these protocol stacks and show how they are used for real-time communication and how real-time applications can be realized over ATM.

7.7 References

1 S. Bradner, ed. *Benchmarking Terminology for Network Interconnection Devices*, IETF BMWG, RFC 1242 (July 1991).

2 P. Ferguson, G. Huston, *Quality of Service*, New York, NY:John Wiley & Sons, 1998.

PART 3

Introduction to Integrated Services

CHAPTER 8

Resource Control on Multiprotocol ATM Networks

8.1 Introduction

Next-generation applications running over IP, ATM, and hybrid networks require features that do not currently exist in either IP or ATM networks or their architectural paradigms. One of the critical problems to be solved in building next-generation multiprotocol ATM networks is combining routing, multicast, and quality-of-service facilities in a manner that makes them accessible at the application level, regardless of the underlying network technology. In addition, these facilities need to be interoperable across mixed networks, allowing computers connected to a legacy network to seamlessly communicate with computers on pure ATM networks.

To address these issues, significant effort has gone into developing and designing technology that can help solve the problems of tying together OSI layer 2 and layer 3 technologies and allowing the bindings to support communication with QoS requirements. To date, these solutions have typically only addressed subsections of the complete problem and have not built comprehensive QoS-capable networks.

Native ATM networks, for example, provide QoS mechanisms via ATM usage parameter control, traffic contracts, and traffic management techniques, but these facilities do not support, nor are they easily translated for, the huge installed IP routing-based or IP multicast-based applications and clients. In addition, IP routing protocols, such as RIP and DVMPR, provide traditional routing and multi-cast capabilities to transport layer applications, but have traditionally lacked support for quality-of-service from the lower layers of the OSI model. Building a complete system, constructed of these high-quality parts, is the goal of integrated services.

In this chapter, we will examine these issues and present past and future methods used to control network resources. We will also illustrate the different traffic and service models that have been developed in the industry and relate them to multiprotocol ATM networks. It is important to understand this information, because it forms the foundation for multiprotocol ATM networks.

The problems faced when building highly scalable routing and multicast protocols have been the subject of intense research for several years. And to a certain degree these technologies can be considered mature, in comparison to the remaining problem of providing QoS guarantees. At the heart of this problem is the industry's desire to provide a means for the application programs to make requests that set aside network resources and, in theory, provide better performance than if the request were not placed. Compounding the desire for application programs to express these needs are the ISP's goals of providing this higher-end, value-added service-level agreement, which could be used as a high-revenue product or a market differentiater.

Granted, there are complex problems associated with providing QoS guarantees by virtue of the agreement that must be made on which parameters need to be negotiated to provide QoS guarantees. These values need to be passed in the QoS negotiation/establishment messages and, at some point, must determine how tight the tolerances should be after the request has been accepted. The solutions regarding these issues have been an area of lively debate, and, consequently, quality-of-service guarantees have been one of the last features to arrive on modern networks.

From a very high level of abstraction, the problems faced in building multiprotocol networks that support QoS can be subdivided into a set of problems to be solved. Consider four tasks that need to be addressed in building a network that will support quality of service and allow a tight integration with a multiprotocol ATM model (see figure 8.1). These tasks, while somewhat obvious, have been the focal point of standards bodies and have required a great deal of industry input to become reality.

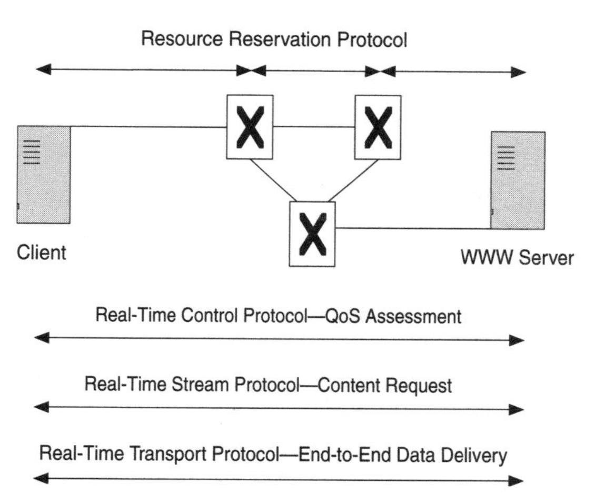

Figure 8.1 Integrated services tasks

- A transport protocol specially designed for real-time applications (provided by the real-time transport protocol) needs to be used. A special protocol would be better than existing ones because the most popular transport protocols today are not optimized for real-time performance.
- A diagnostics protocol used to check the state and QoS (provided by the real-time control protocol). The control protocol is a critical component because it is the feedback mechanism that, like ATM's ABR, will throttle sources and report on QoS being received.

- A method for requesting content such as audio or video clips (provided by the real-time streams protocol.) This protocol can be used to make requests for content on a real-time server.
- A method for requesting different grades of service from the network (provided by the resource reservation protocol). The final component is a signaling protocol that is used by hosts to make QoS reservation requests.

To further the explanation of MPOA and QoS on ATM, this chapter will focus on the integration and philosophy of the above items. Throughout the chapter, emphasis will be placed on the techniques used to request content from servers using standards-based protocols and how client-based applications can manage the resources granted by the network.

We will begin with an overview of the Internet's service model, integrated services. Next, we will present a discussion of its impact on ATM. Building upon the ideas presented, the details of integrating IP and ATM networks into a seamless environment for applications, including support for QoS and traditional IP services, will be presented in the following chapter.

8.2 Integrated services for the Internet

Over the past 20 years, the usefulness and longevity of TCP/IP have been clearly illustrated, and the combined protocol suites have become the core of the Internet. While the underlying technologies, such as Ethernet, ATM, and TCP/IP have changed, these "upper" layers have provided building blocks of historical and technical unparalleled performance and functionality.

TCP/IP's early history can be traced to the ARPANET, where it could be found running over LAN technologies that have long since been replaced with 100Base-X technology. Today, the protocol dominates the multibillion dollar Internet and can be found running over every popular technology: 10 and 100 Mbps Ethernet, FDDI, SMDS, frame relay, ATM, cell switching routers and so forth. This transition in supported protocols is not limited to just physical media types. Due to TCP/IP's performance and functionality, today's popular applications, such as WWW audio and video conferencing, are based on it.

Network designers and protocol developers have recognized the tremendous success of TCP and acknowledged the reasons why this open standard has become so popular. This knowledge has been applied in philosophy to next-generation protocols for QoS and multiprotocols over ATM networks. An attempt has been made to continue this success, and it is this success that drives the continued use of new protocols, and the additional support for new methods of using them in the integrated services for the Internet that are being addressed by the IETF. This work is called integrated services because it focuses on supporting multimedia applications that place QoS demands upon the underlying IP-based network and, in effect, integrates all of today's technology into tomorrow's network.

8.2.1 TCP/IP success

The reasons for TCP's and IP's continued success are important to consider when developing new protocols and network architecture paradigms. First, consider the reasons for success of the OSI transport and network layer protocols and the benefits realized by subdividing a process into discrete components with clearly defined interfaces.

- Delivery of Packets: History has shown that the delivery of packets and the development of packet-switched technologies is a good choice, due to efficient resource sharing and excellent diversity in the range of support applications. The success of packet networks can also be attributed to the physical layer independence that packet switches and datagram delivery pose. As an example of this success, consider that typical IP applications pass data over Ethernet LAN technology, then over FDDI backbone technology, and finally traverses a WAN over Frame Relay or ISDN before reaching the destination. In all these cases, the physical medium changes dramatically; however, the packet format remains a constant. When IP is used in this model, the functionality of packet delivery is delegated to a protocol that supports global addressing but has a problem space limited to datagram transport.
- End-to-end model: TCP, on the other hand, has been very successful, because of its "end-to-end" approach of providing connection-oriented behavior across an IP connectionless medium such as the Internet. Moving error detection and correction to the end systems allows the applications actually sending to decide what information is really necessary for retransmission—thus, building a robust system against network component failure that is not performing undesired error detection and correction. The UDP protocol can be used in the cases where the data are not critical or retransmission would yield stale data. TCP also frees applications from the constraints of determining the correct transmission data rate to use because it implements its own adaptive flow protocol, which will fairly share resources. The sliding window adapts to varying network delays and allows the protocol to reduce the offered load when congestion is detected.

A second reason cited for the success of TCP/IP is the clear division of labor designed into the protocol suite. The lines of division are drawn between delivering a packet and processing it upon delivery in the following manner:

- IP and associated routing protocols provide the means to route packets through the network; however, IP is not involved with the actual delivery to applications and knows nothing about the end application or its requirements.
- TCP takes care of end application needs and issues; therefore, it isolates the end users from the details of the packet-switched network. To an application, TCP provides an error-free bit pipe between the source and destination.

8.3 *Why integrated services?*

During the evolution of Internet applications, from Telnet and FTP to multicast routing and video over IP, it was determined that there was a need for support of new paradigms to extend the function of the Internet. The new functionality is oriented towards making the Internet and intranets reliable for businesses and networks that can support any application imaginable.

The work on integrated services was started because of the demands that new multimedia and Internet/intranet applications place on the networking infrastructure, coupled with the desire to augment Internet technology with function-enhancing tools. Examples of applications that are driving this work include the following:

- Multimedia: distribution of content-rich data containing video and voice and requiring high quality of service
- Collaboration and groupware: data-sharing applications, such as distributed databases, with real-time response
- Distributed simulation: a superset of multimedia application which may also employ multiuser data communication for real-time control over computer simulations
- Entertainment: large-scale distribution of entertainment-grade media that can be categorized at a high priority
- Multiplayer games: entertainment applications and enhancements to distributed simulation.

A key point of the integrated service model is to develop protocols that, in keeping with the Internet's origins, are both flexible and extensible. In addition, as technology progresses, there will be additional applications that will require the integrated services approach and its tools.

The challenge faced by those wishing to use current IP networks for new QoS applications is the question of how to deal with the single-grade, best-effort service category provided by the basic Internet fabric. In addition, even if the network understood how to provide higher grades, how can network managers get their application to negotiate QoS parameters with the network? As a result of the work of vendors and standards bodies, there are now routers capable of supplying multiple levels of performance or service-level guarantees. As will be shown, there are newly developed methods for signaling the need for QoS; however, expanding these capabilities is still in its infancy.

TCP and IP are clearly very good protocols for data, but they do not provide predictable performance on a congested Internet or intranet. Therefore, unmodified TCP/IP is not appropriate for real-time applications if the end users have tight performance constraints and new, overlay protocols are required to realize next-generation applications. In order to provide multiple controlled qualities of service over this legacy Internet, a new architecture has been developed.

The design of the integrated services model that addresses the above issues was a process that spanned three years by the integrated services working group of the IETF. [1, 2] The result of this effort is a set of specifications that divides the functionality of providing controlled QoS between several distinct units. In addition, each different aspect of the QoS supporting infrastructure is subdivided into components, so that there is clear delineation of responsibility. The specification contains details covering the following:

- End-to-end delivery characteristics seen by an individual application's data as those data traverse a network containing at least two elements that support the integrated services model
- Network device behavior realized by applications that utilize nodes conforming to the integrated services model
- Resource reservation dialog rules and languages that are supported by and used between nodes on an integrated services network for exchange of state and control messages
- Performance evaluation criteria potentially used for probabilistic measurements and evaluation of integrated services components.

In the new service model, the supporting mechanisms are typically just software additions made to network hardware. The end computers, for example, will contain new applications capable of generating resource reservation messages and then complying with these reservations while they are transmitting and receiving data.

The interconnecting routers will have new software, allowing them to interpret these messages and segregate traffic into different service classes. They will also need to implement usage parameter control, as described in the previous chapter. However, in some cases, lower-end routers may require additional memory in order to be able to maintain status of resource allocation. In other cases, networks such as shared Ethernet may have difficulty providing integrated service at all. (See figure 8.2.)

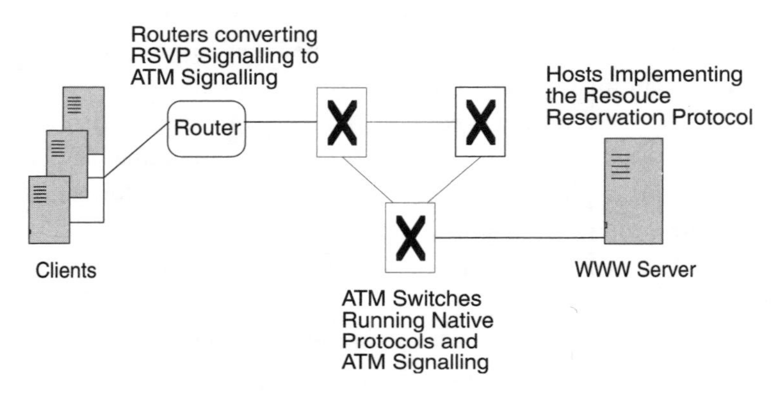

Figure 8.2 Integrated services support mechanisms

It is important to note that the integrated services model is primarily a description of services; the model does not dictate deployment. Because the Internet is made up of heterogeneous parts, which are managed by tens of thousands of different organizations, the end-to-end support of the integrated services model will need time to achieve wide-scale deployment. Therefore, the benefits realized are being seen first on progressive ISP networks usually restricted by administrative boundaries.

8.4 Service models

At the top level, a service model specification describes behavior of individual elements in an integrated services network. In addition to the service specification, the service categories in the following list have been defined to help differentiate what end users can expect to receive from their networks. The integrated services model contains three different degrees of potentially realized qualities of service. These are very similar to the service categories specified by the ATM Forum. Briefly, these can be divided into the following high-level groups:

- Datagram—comparable to today's best-effort Internet service without guarantees, or the ATM Forum's unspecified bit rate service
- Controlled-load—with this traffic model an attempt is made to control the total amount of load, such as ATM's variable bit rate service
- Guaranteed—attempts to provide bounds on load and delay, such as ATM's constant bit rate service.

8.4.1 Datagram service

The datagram service model is the same service that can be expected from traffic that traverses the Internet if the routers have not been modified for integrated services. In this case, the routers provide partial or no guarantees. Most traditional network protocols, such as TCP/IP, supply this type of best-effort service. In a multiservice network, this type of service will be used for commodity communications. If a corporation, for example, builds a high-quality intranet with connections to the Internet, the datagram service would be used for Internet communication, while one of the higher-grade services would be used for internal mission-critical communication. This type of response will clearly not be achieved in the near future, due to the inevitable delays in deploying integrated services technology.

8.4.2 Controlled-load service

The first new service model developed by the integrated services working group is the controlled-load service. This service model was designed for applications that were capable of adaptive behavior—for example, a voice-over IP application, capable of modifying the data being transmitted by changing the compression ratio or realizing voice replay quality.

In the controlled-load service, the network attempts to control the total load of reserved traffic. The network controls the total load by carefully measuring the amount of resources being allocated. Any application that wants to reap the benefits of the reserved resources must also be an active participant in the resource reservation process and the egress traffic shaping process.

The controlled-load service does not provide any quantitative performance assurance to the traffic being carried. There are no guarantees made with respect to the maximum delay or congestion that may be experienced. Instead, this service class attempts to impose no substantial requirements on the network elements and simply attempts to accurately predict what the realized QoS will incur when the resources are shared intelligently.

The only assurance with controlled-load service is that performance will be as good as an unloaded datagram network. Granted, this performance guarantee, while not stellar, may prove to be more than adequate for the vast majority of applications on the Internet or intranets that desire higher grades of service. This is due to the fact that traditional IP routers and legacy networks provide very good performance when unloaded; therefore, many protocol/networking experts in the industry feel that this may be the only protocol necessary in the new integrated service model.

A final point important to consider with respect to the controlled-load service is that it is widely believed to be the dominant real-time service in initial deployment of integrated service support. The controlled-load service is a simple protocol for programmers to understand and to which end users can characterize their applications. An additional benefit includes the potential for functioning well in networks with only partial deployment, because the performance of interconnecting links can be predicted with a degree of certainty.

8.4.3 Guaranteed service

The final model developed by the IETF's integrated services working group is the guaranteed service class. This service class is designed to provide applications with the tools to communicate with a network capable of understanding reservation requests. It provides a service supporting firm bounds on data throughput and delay along the path between communication end stations.

Guaranteed services provide quality guarantees that can be explicitly stated in a deterministic or statistical representation. Deterministic bounds are specified by a single value, such as average bandwidth or peak required bandwidth. Statistical bounds are determined by a statistical measure, such as the probability of errors, while predictable services are based on past network behavior. QoS parameters for reliable services are estimates of consistent behavior, based on measurements of past behavior.

In order to achieve this highest degree of service-level guarantee, the service class imposes substantial requirements on network elements.

- Every element along the path must provide delay bounds and maintain conformity with the advertised bounds.
- Topology changes must force the recompilation of the delay bounds and subsequent new advertisement messages sent to end stations notifying them that their realized quality may have dramatically changed.

The guaranteed service class is intended to provide the highest achievable service in the current version of the integrated services model. This class is believed to provide a mathematically provable delay bound and packet loss for a given path when the resources reserved utilize guaranteed service class definitions. Because this protocol places tight constraints on network components and their ability to advertise correctly and then supply the resource, we will discuss some examples where it may not be usable.

It may be that the guaranteed service protocol will not be capable of being implemented if the intermediate network systems do not support the class. This may also be the case if data communication is traversing a shared media, such as Ethernet. These devices and legacy networks are potentially unacceptable to the guaranteed service integrated services model, because they can provide a large degree of variance in the realized QoS by end station applications. These concerns further fuel the belief that the controlled load model will not be very successful. Only practical experience will prove the success of guaranteed service.

8.5 Integrated services model components

In defining the integrated services model, the IETF has subdivided the tasks necessary for providing QoS support into various categories based on function. The division has been useful from the standpoint that it allowed the problem to be better understood by novices to the industry, and it also allowed the systems to follow Internet standard practices. This means that the end system can be built from smaller components that are scalable and flexible. The components of the integrated services model are:

- Resource reservation protocol—used to set up and control state between end stations
- Admission control—used to control the total amount of load the network accepts
- Packet classifier—used to prioritize packets into different qualities
- Packet scheduler—used to intelligently control the introduction of new traffic into the network.

The interworking between the various components of the integrated services model is shown in figure 8.3.

As described in the preceding chapter, there is currently a need for integrated real-time service support that includes IP routing, ATM style QoS, and multicast features. These services should be available to applications running on ATM networks, IP networks, IP over ATM networks, and mixed networks. The integrated IP-ATM network model is shown in figure 8.3. In this model, there are several possible communication paths between hosts:

- ATM host to ATM host

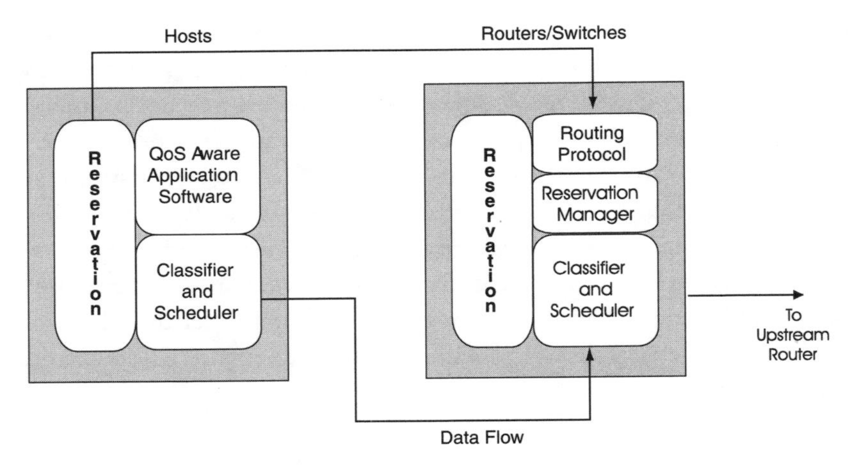

Figure 8.3 Internetworking of integrated services component

- IP host to IP host
- ATM host to ATM host over a non-ATM IP network
- IP host to IP host over an ATM network
- ATM host to IP host.

To enable an integrated IP/ATM network, the features of the ATM model, the IP model, and QoS must be interconnected within the network and available at the application level. The goal of the integrated services model is to mask the underlying technology from the application, while still providing the following features:

- Internetwork routing allows applications to achieve their desired performance from the network via optimal path section.
- Multi-cast capability allows one-to-many, or many-to-many communication flows.
- QoS facilities are parameters that describe the desired characteristics applications can expect from the network and the mechanisms used to request these facilities.

The coupling of ATM to IP was a process that had many steps in its evolution. Basic IP over ATM communication was first enabled through the development of the IETF's Classical IP over ATM (RFC 1577) and the ATM Forum's LAN Emulation specification. These protocols took an early look at how to move data over an ATM network while still maintaining compatibility with legacy networks.

LANE's approach was to emulate the functionality of either an Ethernet or a Token Ring network by establishing servers to process broadcast packets. Classical IP Over

ATM's approach was to mimic the ARP service with an ATM-attached ARP server. One of the major drawbacks to both of these solutions was the lack of QoS support. Both of these standards are important because they provide the underpinnings of an MPOA system. They offer this function by providing the protocol that is used to move data between ATM LANs.

The protocols described in LANE and Classical IP, however, are just the starting point for an integrated services model that supports ATM. The path that the ATM Forum has chosen for the next set was to construct a new protocol, Multiprotocol over ATM (MPOA), that would evolve LANE into a service with QoS support.

Like LANE, MPOA is being released in several versions. The first version addressed virtual circuit setup policies that utilize LANE, coupled with more advanced protocols for inter-LAN communication across ATM switches. All of these phase 1 activities are designed to be relatively simple and quickly implementable; therefore, they are built to run on networks without QoS requirements. Once the foundation of the version 1 implementation has been built, network designers can expand upon this to supply IETF's integrated services–based applications with hooks to ATM's QoS and multicast capabilities.

8.6 Integrated Services Evolution

The evolution of the integrated services model has been a long iterative process for over four years. In that time the work completed in the IETF and heavily augmented at the University of California, Southern California, has led to the development of the integrated services core components:

- Framework—illustrating the scope of work
- Service definition—describing the benefits a user of the integrated services model can expect
- Component requirement—describing the roles of each device in an integrated services network.

The first generation of the integrated services working group's finalized specification became available at the end of 1996 with compliant products arriving on the market shortly thereafter. The first generation of products has produced a set of standards that allowed network designers and managers to implement systems supporting QoS over frame-based IP networks. Granted, the support for QoS was limited to their administrative boundaries, because of the reluctance of ISPs to support the protocol beyond small pockets in their networks. While these first-phase implementations and specifications were limited by geography, due to lack of ISP support, they are nevertheless critical for corporate and campus intranets.

The next phase of the integrated services evolution will take several paths. These can be subdivided into the following categories:

- Prototyping and knowledge building
- Missionary base building
- Interworking with legacy/competing systems
- Acceptable use policy development
- Accounting and measurements.

The first path is acquisition of knowledge gained by network implementers as they roll out support for these services. There is still a great deal to be learned about the interaction between the components in real-world global Internet. The second path falls on the shoulders of application software developers. There is a need in the industry for next-generation software that can make use of the integrated services protocols. A tremendous benefit to developers is the fact that software developed for the integrated services model will work on any QoS-aware network. In addition, a new software market will be created by this, enabling technology for applications from virtual private networking to high-quality multimedia distribution.

The next problem to be solved, interworking with legacy and competing systems, may be one of the most difficult to resolve. Interworking with legacy systems will test the ability of QoS to truly be supported over packet-switched networks and prove that the technology can compete head to head with cell switching. In addition, there are a number of technical problems (e.g., interworking with ATM's signaling) that have been discovered by the standards bodies; these are still in the early stages of field trials.

The last path of the integrated services work will involve analysis of administrative, financial, and policy mechanisms for global deployment. These may be the most challenging problems faced by the integrated services supporters, as they begin to grapple with the issues of end-to-end support for their new protocol.

Thus far, we have described the ATM Forum's and IETF's views of QoS service models. In the next three chapters we will describe the technology used to realize these models and how legacy networks can be migrated towards QoS-capable systems running in an MPOA environment.

8.7 References

1. H. Schulzrinne, S. Casner, R. Frederick, V. Jacobsen, *RTP: A Transport Protocol for Real-Time Application*, RFC 1890, January 1996.
2. H. Schulzrinne, *RTP Profile for Audio and Video Conferences with Minimal Control*, RFC 1890, January 1996.

CHAPTER 9

Real-Time Transport and Messaging Protocols

The preceding chapters have reviewed some of the facilities available to an MPOA system, the goals of the integrated services working group, and some techniques for achieving these goals. In this chapter, we will begin to describe how a complete MPOA system can be built. Our focus now will be on protocols that have been developed to support integrated services in both a legacy LAN and an MPOA system. This material is important, because it describes how computers can exchange real-time data. From a high level of abstraction, the protocols described in this chapter can be used in any sophisticated network, and MPOA is just another method of providing the basic means for the computer to communicate.

Protocol designers active in the IETF and the ATM Forum over the last three years have expended considerable effort on real-time protocols. These protocols are called real-time, because they are used when there are tight constraints on the quality of service that must be delivered from the network—for example, the total transit delay or interpacket arrival time must be bounded. There are three primary protocols that have been developed to support real-time quality-of-service data over IP and a fourth, related protocol used to stream multimedia content.

1. The Real-Time Transport Protocol (RTP) is a real-time end-to-end protocol utilizing existing transport layers for real-time applications.
2. The Real-Time Control Protocol (RTCP) provides feedback on the quality of the data transmitted, so that modifications can be made.
3. The Resource Reservation Setup Protocol (RSVP) is a multicast capable resource setup (signaling) protocol primarily designed for IP. RSVP is a general-purpose signaling protocol and could be used to map resource reservations to ATM signaling messages.
4. The Real-Time Streaming Protocol (RTSP) is a transport layer protocol designed specifically for controlling the transmission of audio/video over the Internet.

When used in an MPOA system these protocols would be used to carry real-time data encapsulated inside ATM cells. The remainder of this chapter examines how RTP, RTCP, and RTSP are designed, internetworked, and deployed. RSVP, and its interworking with ATM, is discussed in chapter 10.

9.1 Real-Time Transport Protocol: RTP

When developing protocols that would be used first to allocate resources and subsequently to carry real-time traffic across intranets and the Internet, IETF and the ATM Forum working groups decided that it would be best to start with a clean slate. The first protocol described in this chapter is the Real-Time Transport Protocol (RTP), as specified in RFC 1889 [1].

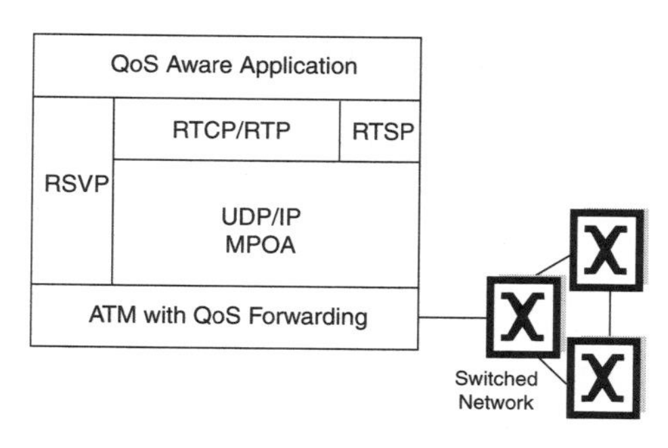

Figure 9.1 Protocol interaction

RTP's primary function is to act as an improved interface between real-time applications and existing transport layer protocols, but not necessarily those providing the connection-oriented behavior of TCP—that is, RTP does not dictate which transport layer protocol is used. It provides functions that allow transport protocols to work in a real-time environment (see figure 9.1). One of RTP's chief design goals was to provide a simple, scalable protocol independent from the underlying transport and network layers. The desire for good scalability is driven not only by the large scope of the Internet but also by the desire for good interworking with multicast IP.

In providing functionality just above the transport layer, RTP is designed to provide end-to-end delivery services for temporally sensitive data with support for both unicast and multicast delivery. The underlying network is assumed to be any packet-switched network in which a packet will arrive at its destination with some certainty. Due to the nature of packet switching, variable delay is to be expected. Additionally, due to packet switching and routing, packets may arrive out of order.

It is important to keep in mind that RTP is only a transport protocol; it supplies similar functionality to UDP and, in some cases, can be carried inside a UDP payload. RTP provides the following features and functions:

- Data source and payload type identification, which is used to determine payload contents
- Data packet sequencing used to confirm correct ordering at the receiver.
- Timing and synchronization, which is used to set timing at the receiver during content playback
- Delivery monitoring, which facilitates transmission problem diagnosis or provides feedback to the sending computer on the quality of data transmission
- Integration of heterogeneous traffic, which is used to merge multiple transmitting sources into a single flow.

RTP does not provide any quality-of-service guarantees or deliver data reliably; it is a protocol that monitors and helps control the flow from transmitter to receiver. The functions of quality-of-service guarantees and delivery are the responsibility of RSVP and the packet-switched network-supported QoS.

The suite of protocols also contains definitions regarding which component should perform which specified function. The RTP component carries individual real-time data streams with a source identifier, payload type, time, and sequencing information. The feedback component, which is covered in greater detail in the next section, monitors application performance and conveys information about the session (e.g., information about participants).

9.1.1 RTP usage scenarios

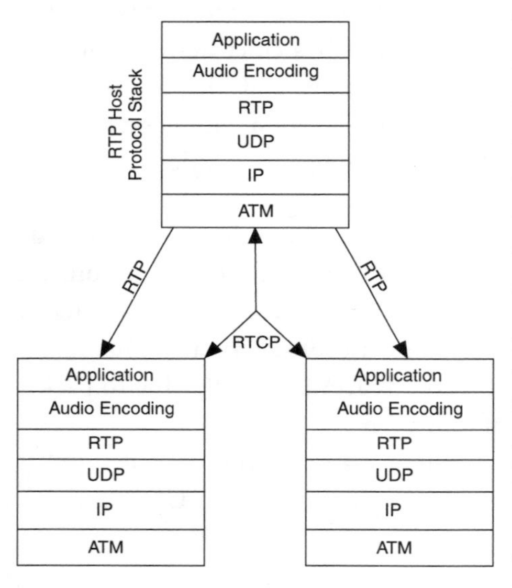

Figure 9.2 Using RTP

To illustrate the interworking between RTP and RTCP, it will be helpful to consider an example of an audio conference session. The example system is shown in figure 9.2. The diagram illustrates the playback of an audio-only multimedia system. Packets on the left arrive at the host, where they can be buffered and examined prior to fully decoding the payload and reproducing the sound on the host computer's speakers. As with other RTP applications, receiver feedback and group membership information is provided via RTCP.

When packets arrive at the destination, each carries a sequence number and time-stamp. The sequence number is examined to determine the correct sequencing of data and also to record the percentage of lost frames. The RTP packet's timestamp is used to determine the interpacket gap in the following manner: The timestamp value is set by the source as it encodes the data and transmits the packet into the network. As packets arrive at the destination, the change in interpacket gap can be examined, and then during playback this information can be used to regenerate the contents at the same rate they were encoded. By utilizing buffering at the receiver, the source can attempt to pace the traffic independent of the jitter introduced by the packet-switched network.

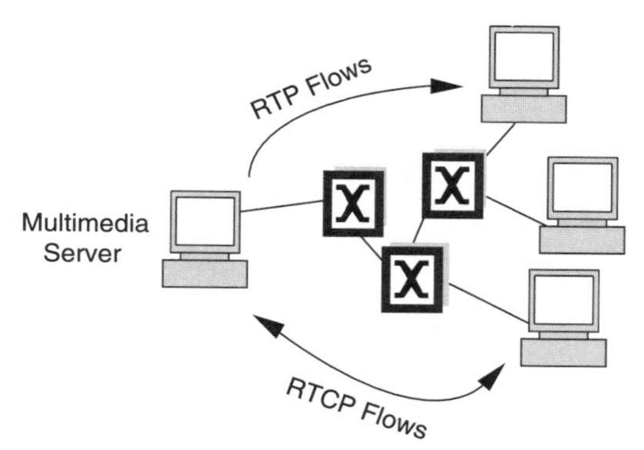

Figure 9.3 Using RTP with audio and video

The concept of source synchronization can be extended to sessions that contain multiple concurrent types of content—for example, a conference with multiple audio and video sessions. In this type of application, the recommendation is that each type of content be carried over its own unique RTP session. When the RTP session arrives at the destination, the different sources can have their synchronization source fields compared to synchronize the audio with the video (see figure 9.3). The RTCP facilitates the synchronization of multiple RTP sessions and reports on their quality. Just how RTCP provides these functions is addressed in the following section.

An additional example regarding the use of RTP is with HTTP multicast. In this example, the content would be generated from possibly the multicast of WWW server pages (used for proxy updates) or network news distribution. Reliable multicasting is beyond the scope of this book.

9.1.2 RTP format and terminology

An RTP session is defined as communication between hosts, identified by a pair of transport addresses along with the port number assigned to RTP. The fields and format of an RTP packet are shown in figure 9.4. Several of the fields are unique to RTP and deserve some explanation in light of the RTP terminology.

There are two fields used to help identify the stream, Synchronization Source (SSRC) and Contributing Source (CSRC). The SSRC is a 32-bit identifier generated by the source of a single real-time stream. The CSRC is a 32-bit identifier formed by a source that contributes to the combined stream produced by an RTP mixer. A mixer in this case is a device that physically combines several streams into one super RTP flow. The CSRC count is used to numerate the participants when multiple sources are being combined.

The first 12 octets are present in every RTP packet, while the list of CSRC identifiers usually will not be present, since it is used solely when broadcasting an RTP session. The complete RTP header contains the following fields:

- A version used to identify the release of RTP
- A padding-bit which specifies that the payload has additional padding at the end

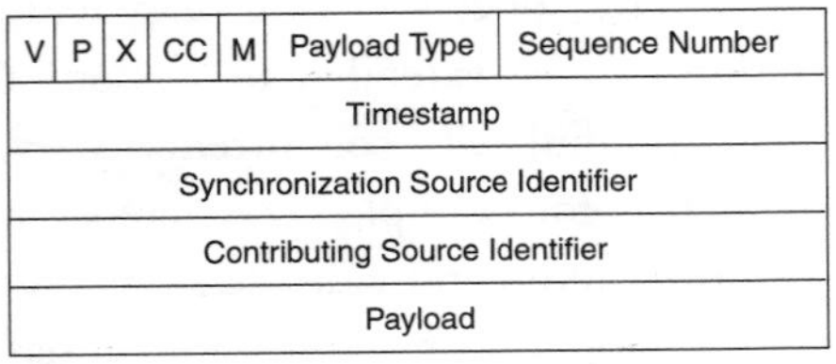

V	P	X	CC	M	Payload Type	Sequence Number
Timestamp						
Synchronization Source Identifier						
Contributing Source Identifier						
Payload						

V — Version
P — Padding Indicator
X — Extension Indicator
M — Market Bit
Payload Type
Sequence Number
SSRC — 32-Bit Source Stream Identifier
CSRC — 32-Bit Value Used to Identify Contributing Sources in Mixed Stream

Figure 9.4 The RTP header

- A counter for the number of CSRCs that follow the fixed header
- A payload type identifier, which specifies the type of data being carried in the RTP packet
- A sequence number which increments to count the number of packets in a stream
- A timestamp, which denotes the instant the first octet in the RTP data packet was sampled
- An SSRC field, which identifies the synchronization source associated with the data (this identifier is chosen with the intent that no two sources within the same RTP session will have the same value)
- A CSRC field with a list of objects that identifies the contributing sources for the data contained in this packet (the total number of identifiers is given by the third field).

The second possible format for the RTP header is a "compressed" version. The compressed header is used when only one source is contributing to the payload, as would be the case in a nonmulticast session. The RTP packets are marked as compressed when the first bit of the header, the version bit, is set to zero. When the version is zero, the header will be modified to only contain the fields up to and including the SSRC.

9.2 Real-Time Control Protocol: RTCP overview

As described in the preceding section, RTP is a very simple protocol designed to carry real-time traffic and provide a few additional services that are not present in the existing transport protocols such as UDP. With RTP, receivers can utilize the timestamp, along with sequence numbers, to better synchronize sessions and improve playback. To complement RTP, the IETF has designed a protocol, RTCP, which is used to communicate between the sources and the destination. This protocol is not used to establish QoS parameters with the ATM switch; instead, it is oriented towards client/server state exposure.

The design goals of RTCP are to expand this system by providing the following features:

- Feedback on the quality of the data distribution
- Persistent transport level identifier for RTP sources
- Automatic adjustment to control overhead
- Session control information to maintain high-quality sessions.

Therefore, RTCP should be considered as a primary technique in which hosts using RTP can communicate "out of band" to exchange information about their state. The information exchanged would be used to identify the quality of service being delivered by the network. RTCP can illustrate a large degree of packet loss, the value of the roundtrip time, or whether the network is providing high jitter values.

9.2.1 RTCP reporting

In order to achieve these goals, the RTCP utilizes the exchange of packets that express the end point's state. There are two main packet types exchanged: Sender Reports (SR) and Receiver Reports (RR). Each of these reports is accomplished by generating a packet containing fields that describe the state of the session. In addition, several report packets can be concatenated together to form a compound RTCP packet, which is transmitted over the transport layer as one packet.

Each of the SR and RR packets contains the SSRC of the sender. In the case of an RR, it contains the SSRC of the first source. Perhaps more important for maintaining a high quality of service, both packet types contain fields that can be analyzed to determine the percent of successfully received data packets. The fields used to help determine the quality of service are as follows:

- The fraction lost versus the total number of packets, which provides an instantaneous feel for the percentage of loss
- Cumulative number of packets lost since session began
- Highest sequence number received, which reveals how much of the current data in transit has been received.

- Interarrival jitter, which can help specify how much buffering is being used at the destination to faithfully replay the source; the jitter value at the time of sending the report
- Last SR timestamp received, which reports the time when the last SR was received
- Delay since last SR timestamp.

9.2.2 Analysis of sender/receiver reports

Due to the large amount of information provided by the RTCP packets, there are several ways that these messages can be interpreted. Throughput can be determined by comparing the number of packets and bytes transmitted since the last report. Roundtrip times between the sender and the receiver can be calculated by subtracting the arrival time from the sum of the timestamp of the last report and the delay since the last report. Finally, packet losses are determined by analyzing the fraction lost since the previous SR or RR packet sent, or the total number of RTP data packets lost.

In the case of a multicast session, the RTCP packets can be collected and analyzed to determine temporal state—for example, the jitter values can determine transient congestion versus the loss values that provide persistent congestion indication. A comparison of loss values can determine the size of the subset of total recipients experiencing congestion and if the problem is global or local. With this information, the source can decide if the sender should make any adjustments to maximize available bandwidth.

Additional RTCP packet types support the ability to query the source for descriptive information, indicate the end of a session, or support application-specific functions via the SDES, BYE, and APP packet types, respectively.

9.3 Real-Time Streaming Protocol: RTSP

The final protocol covered in this chapter is the Real-Time Streaming Protocol, or RTSP, which is an OSI application layer protocol designed to supply a mechanism for making requests about the delivery of real-time content. In a real-time multimedia system, the intent is to use RTP to deliver content and then use RTCP to determine the quality of service being provided. With these two protocols in place, there are still two critical components missing: first, a means of notifying the network of the required bandwidth of the upcoming transmission (the topic of the next chapter) and, second, a means for requesting content from a multimedia server.

There are three different formats content could assume in which the RTSP is used to request transmission:

1. Real-time stored clips, which will include the set of prerecorded multimedia stored digitally on a server
2. Non-real-time stored clips, which will include content typically transmitted via HTTP or MIME
3. Real-time live feeds, which will be feed from radio or televised content digitally converted just before transmission.

In much the same way that traditional client/server systems are architected, the RTSP clients generate messages that are transmitted to their multimedia content server. RTSP does not carry data that are actually appreciated or viewed by the end user; it only provides the functionality of a signaling and control protocol, giving random access to content.

The design requirements of RTSP are as follows:

- Supply a means for requesting real-time content
- Have a means for content playback control (e.g., pause, start, stop)
- Provide a technique for starting playback at any point in the content (e.g., play the second track on an audio CD)
- Secure a means to request information about specifics on the content
- Allow the protocol to specify the transport layer protocol selection.

RTSP uses TCP when exchanging its control message between the client and server. The protocol has been designed to work on both a multicast and unicast environment. When exchanging signaling messages with a server, the client has the option of using the Session Control Protocol (SCP) session to improve efficiency.

After the multimedia server has received an RTSP request for content, it then transmits the requested data, most likely over RTP, to the client. If the content consists of several components, for example—voice and video—then potentially multiple streams will be created between the server and client. When multiple streams are created, each one will be used to carry a portion of the data: one stream for video and another stream for the audio.

The RTSP operates by exchanging signaling messages between the client and the server following a query/response mode. There are three main categories of RTSP messages:

- Global control is used to govern all sessions between the client and server.
- Connection control is used to establish, maintain, and terminate individual content data streams.

- Custom control is used to provide exception messaging beyond the scope of connection control.

From a high level of abstraction, the protocol operates when the client registers itself with the server, which is a global controller. This step is then followed by requests for data streams which are the connection control messages. Therefore, the first sets of messages that must be exchanged before any multimedia data can be transmitted are the global control messages. The possible global control messages are as follows:

- `HELLO`—registration request
- `GOODBYE`—global session termination message
- `IDENTIFY`—a server's request for authentication
- `IDENTIFY REPLY`—the client's authentication response
- `REDIRECT`—point to other content servers
- `OPTIONS`—indicates miscellaneous functionality.

The registration process begins with the client sending a `HELLO` message to the server. The message contains the client's RTSP version information and may contain the client's host name, machine type, operating systems, and so forth. Upon receiving the `HELLO`, the server then replies with a similar `HELLO` message, but the body of the message contains the server's `RTSP` version number. From that point onward, the client and server can communicate using the lower of the two version numbers.

It should also be noted that this communication takes place over `TCP` therefore, the exchange of messages is deemed reliable. It is also possible for the server to redirect connection requests by using the `REDIRECT` message. The `REDIRECT` message uses a URL, which identifies the new location for the client to access content.

The next phase of the global initialization process involves the option of authentication. Either the server or the client can generate an identification (ID) message, which requests the recipient to prove its identity. The actual technique used for authentication is left to the discretion of the hosts and may range from a simple password check to a challenge/response mechanism. If the response to the ID message is not acceptable, then the TCP connection is closed immediately.

Termination of a global session is accomplished with a `GOODBYE` message. Clients only send these messages as they are gracefully closing their TCP sessions. The final global message type is the `OPTION` message. It was designed to provide the ability to incrementally improve the RTSP.

9.3.1 Connection control messages

Once the global message exchange has taken place and the client and server have authenticated each other, the next phase, content request, can take place. The process of requesting audio and video is rather straightforward: The client specifies the objects that interest it and the parameters used during transmission. The message types used to request content are as follows:

- `FETCH`
- `STREAM HEADER`
- `SET TRANSPORT`
- `SET SPEED`
- `PLAY RANGE`
- `STREAM SYNC`
- `SET BLOCK SIZE`
- `STOP`
- `RESUME`
- `SEND REPORT`
- `RESEND`

Even though there are several possible messages, the flow of a request is very predictable. The process starts when a user makes a request to access some content on the server. This request could come, for example, from a user who comes across an interesting video clip while operating a WWW browser. The request for the video is carried in the `FETCH` message. The `FETCH` contains a pointer to the video clip in the form of a URL. It also contains the client's estimate of the bandwidth available for the stream.

If the server accepts the request from the client, it then responds with a `STREAM_HEADER` message. If a problem was encountered, then an error message is returned. The `STREAM_HEADER` message specifies the exact bit rate of the stream that will be the product of the bandwidth requested in the `FETCH`. It also contains a great deal of information about the video clip—for example, the STREAM_HEADER message will specify the following:

- Last modification date of the content
- Length in milliseconds
- Maximum packet size to be used during the transmission

and:

- If the content is live or prerecorded
- If the content can be cached at a WWW proxy

- If the content can be multicast
- If the content can be carried via UDP.

When the client has received the STREAM_HEADER message, the server is in a ready state. The client need only notify the server to begin transmitting the stream. Before the client issues the request to begin playing the video clip, it has the option of placing a network resource request. This is accomplished with the SET_TRANSPORT message, which acts as the interface between the application and the RTP protocol. By setting the transport mechanism, the client has the ability to specify a multicast session and the UDP port number for the data stream. The client can also optionally reduce the maximum packet size with the SET_BLOCK_SIZE message.

The next step is to issue a command to actually begin sending the data via the parameters set by the SET_TRANSPORT message. The request to send is done by the PLAY_RANGE message. Its parameters are just the starting and ending points, specified in milliseconds. In response to the PLAY_RANGE message, the server will positively acknowledge and begin sending the request with a STEAM_SYNC message which contains the beginning sequence number and timestamp for the upcoming video.

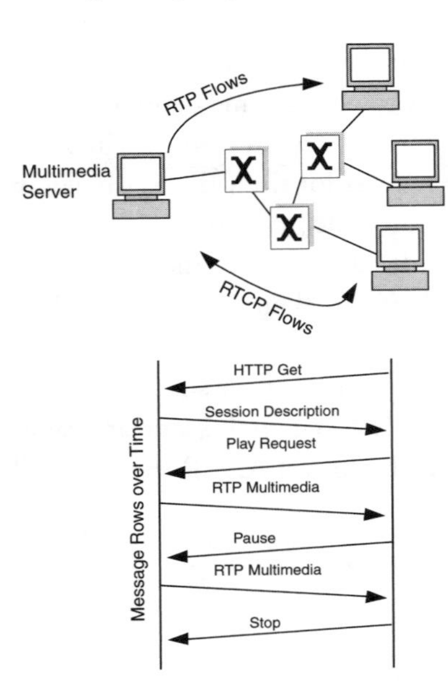

Figure 9.5 Example use of RTSP messages

While the video is in play, the client has the ability to control the data stream in much the same way as playing a movie on a VCR. The data stream can be paused or terminated with a STOP message. It can also be restarted with a RESUME message.

A diagram illustrating the flow of the RTSP message exchange is shown in figure 9.5. In the figure, a client is requesting a video file from the server. The client first locates the server with a HELLO and is then asked to authenticate itself. After authentication, the client issues a FETCH containing a URL to request the video clip.

Finally, there are two diagnostic messages that can be exchanged. The server can issue a SEND_REPORT to the client to help identify the quality of the reception. There are reply options for a client who has received this message. At a minimum, it must generate a PING message directed at the server. If possible, the client can generate and reply with a detailed report containing such items as the number of packets received and internal buffer utilization.

The other diagnostic message is the `UDP_RESEND`, which is used by the client to request the retransmission of data. The message has fields allowing the client to specify exactly what sequence number to begin with and how many subsequent packets need to be retransmitted.

9.3.2 Custom control and protocol extension messages

Another class of messages supported by RTSP is used to build general purpose client/server information exchanges. A typical use of this feature occurs when the server wishes to direct the client to change to another server for acquisition of additional multimedia material. Another possible use would be when the server wishes to advertise the availability of a new service—for example, the nightly news video program stream has arrived and is ready for playback. The two message types used to redirect or announce new services are `GotoURL` and `NewURL`, respectively.

9.4 Summary of RTP, RTCP, and RTSP features

This chapter has focused on the protocols used to request real-time content, transport it across a QoS empowered network, and then monitor it to realize QoS at the receiver. The RTP has been developed with flexibility and scalability in mind and is now being used as the core protocol for real-time transport on both pure IP network and hybrid MPOA systems. It provides transport of data with an inherent concept of time, and it also provides an excellent means for transmitting real-time data, because it, unlike legacy transport layer protocol, has been optimized for that task. RTP packets include only enough information to provide real-time support: a source identifier, sequence number, timestamp and payload type. With this simple header, the RTP protocol is lucid and capable of high-speed data handling.

The counterpart to RTP is RTCP. RTCP provides basic session and monitoring features, client/server communication, and hooks for application-specific control information. With RTCP applications, developers and end users have ready access to a query/response mechanism which will allow them to determine, in real time, what QoS the network is actually delivering.

Finally, the RTSP performs the key function of content request and control. With RTSP, developers have a consistent interface to multimedia and other real-time content. Similar to traditional client/server systems, the RTSP client generates messages directed to the multimedia content server. The server can be a generic WWW server, enhanced to understand RTSP and thus allow the WWW browser to request content and control random access to the content.

In the next chapter, we will examine the signaling protocols used for making network QoS requests and how these requests are mapped to sophisticated switch technology such as ATM.

9.5 References

1 H. Schulzrinne, S. Casner, R. Frederick, and V. Jacobson, *RTP: A Transport Protocol for Real-Time Applications.* RFC 1889, GMD Fokus, January 1996.

CHAPTER 10

Reserving Resources and QoS on MPOA Networks

The previous chapters have described architectures and protocols that ISPs are using to build their networks, and in this chapter we examine in greater detail how these new networks can be used to provide differentiated qualities of servers. This information is critical for ISPs and corporate network managers, because it allows them to provide two of the key missing components on today's Internet: reliability and predictability. When the network has been built to provide quality-of-service guarantees, businesses can begin to think of the Internet as a utility rather than a novelty. The final component required before the Internet reaches this utility state is security.

The ability of an ISP to provide quality-of-service guarantees can be achieved in two ways. First, and simplest, the ISPs can statically provision their networks to provide priority to packets of a certain type. This would be used, for example, to ensure that email was carried over congested Internet links while regular WWW traffic was dropped. This method of providing quality-of-service guarantees is simplest because the ISP's routers are configured in advance to recognize the bit patterns of different types of Internet data. However, with this simplicity there is a serious tradeoff in flexibility; therefore, an alternative method has been developed, which allows for any program to request a higher priority at any time. This second technique uses a protocol known as the Reservation Setup Protocol (RSVP), which moves the real-time provisioning over quality-of-service away from the ISP and back into the hands of the end users.

This chapter addresses the Reservation Setup Protocol used to request state in ISP routers and reserve services on behalf of end-user applications. The RSVP signaling protocol can be considered as a network controlling protocol for real-time services in connectionless networks. Briefly, RSVP is a signaling protocol that utilizes traditional IP networks, which may be interconnected with ATM switches running MPOA protocols, to carry its resource reservation messages from sources to destinations. Along the path between the source and destination, the resource requests are used to gain permission from admission control software regarding the availability of the desired resources. Then, resource requests reestablish the reservation state, essentially securing the reservation. When the desired request cannot be fulfilled, a request failure message is generated.

This point is a key marketing differentiater for ISPs, because it allows them to provision a network that has a single physical connection to their customer: on that connection they can provide dynamic quality-of-service. The user of an RSVP-capable ISP network will have several choices afforded to them when transmitting IP packets across the Internet. The default, lowest-grade service will be the same best-effort Internet traffic that has been available for years. But now the ISP is able to determine, via the RSVP signaling message, how much bandwidth each customer requires, what latency is requested, and what type of service the customer expects. Of course, the ISP will now charge a higher price for its utility-grade service.

The focus of this chapter will be the protocols used to realize a value-added ISP network—namely, how end users can request quality of service from a network and, integrate IP and next generation ATM/MPOA networks into a seamless environment for applications supporting differentiated quality-of-service and traditional IP services.

While still in its infancy, it is clear that RSVP will provide a critical means for network managers to provide differentiated QoS that can be metered and subsequently billed. The protocol will follow an evolution similar to that experienced by ATM: moving from the laboratory into research and education facilities and followed by a migration into mainstream business applications. It is also believed that RSVP will face tough criticism—some justified and some not—as did ATM. There was a time when it was fashionable to berate ATM, and we will no doubt see this same stage with RSVP. However, as corporate intranets and the Internet evolve towards becoming utilities similar to telephony, the resources that RSVP brings to the table will be indispensable.

10.1 RSVP's history

RSVP was conceived in 1991, while the first generation of Internet-based multimedia tools was being deployed at Lawrence Berkeley National Laboratories (LBNL) and Xerox's Palo Alto Research Center (PARC). Researchers at LBNL were working on tools that would permit workstations to transmit voice and video conferences over IP-routed networks interconnecting their campuses. At the same time, research was underway to develop a scalable solution for IP multicast at PARC. When researchers wanted to test their new applications on the network, they manually provisioned the interconnecting routers and allocated resources via a human operator. These early tests proved difficult to manage and clearly demonstrated the need for automation to ensure the success of large-scale, real-time applications.

When designing a suitable resource reservation protocol for large-scale deployment, the researchers at LBNL and PARC soon realized that they must establish a set of baseline requirements. As with most of their work developing Internet protocols, one of the chief requirements was to have the protocol use Internet resources efficiently and, more importantly, scale globally. In addition, it needed to support the ability for multicast (i.e., one transmitter with multiple receivers) applications smoothly in a heterogeneous environment. Finally, the protocol had to adhere to TCP/IP's design principles of being robust enough to handle packet loss and adapt well to changes in network topology.

After the default IP data path is set up by either traditional routers or ATM switches, RSVP is used to deliver QoS requests to each switch or router along the path, an often overlooked key point. Thus, it is a signaling protocol, not a routing protocol. Each node along the data path processes the QoS request, possibly reserving resources for the connection, and then forwards the request to the next internetworking device along the selected path.

RSVP is a very robust protocol and is designed to support multicast and unicast data delivery in a heterogeneous network environment, however, it is clearly designed with IP in mind. In addition, the flow of IP datagrams produces a unidirectional reservation—that is, the end stations are only specifying resource reservations for one direction at a time. Therefore, two reservation requests are needed if bidirectional quality of service is desired.

The protocol uses IP datagram and the basic method of carrying the signaling messages. This allows the RSVP messages to be transported over any ISP's network. The message can be passed from router to router and only processed by routers that understand RSVP. In the case where the packets cross non-RSVP-capable routers, the messages are ignored.

Because the RSVP signaling messages are being carried in IP packets, there are several methods for placing the RSVP data into the IP payload. The end stations that generate the RSVP messages have the capability of transmitting the messages in "raw" (i.e., directly mapped into IP packets) mode, using TCP or UDP encapsulation. If transmitted in raw mode, the RSVP message is placed inside the IP payload with a protocol type of 46. The UDP method, which at this point is the most common encapsulation found on host implementations, is supported for systems that cannot generate raw IP packets—for example, some PC IP software stacks lack a UDP implementation.

10.1.1 Soft versus hard state

Coupled with the desire to develop a new protocol, and the evolutionary steps required in the Internet, the RSVP suite forces little permanent state upon the network devices supporting the protocol*. This state is referred to as 'soft,' because it is believed that handling dynamic routing changes should be a normal procedure, not an exception; therefore, routers should be constantly updating their reservations when they periodically receive resource requests. This approach is similar to the philosophy used by the early Internet architects, in which a design goal or consideration involved datagram-routed networks providing only "soft" routing state; therefore, they could be changed at any time. When this idea is applied to RSVP, the product is a protocol that expects routing

---

* The amount of state maintained for each reservation is actually a point of contention between the protocol designers and implementers. Some feel that early estimates grossly underestimated the amount of memory required in routers to support RSVP. Currently ballpark numbers show the amount of memory to be approximately 1KB/reservation.

changes and reacts well to them. It is also a protocol that is independent of the Internet's mature routing protocols.

In order for soft state to work, the system must be periodically refreshed with the desired state. When using RSVP, a signaling system is developed where resource requests are made and then periodically refreshed. The refresh messages are identical to the original resource request messages, only repeated. If the path from source to destination has changed, due possibly to routing change or link failure, then the next refresh message will create a new resource reservation. Or, in the worst-case scenario, the new network route returns an error message specifying that the requested resources are not available.

Dynamically changing routes pose some interesting problems to quality of service. If a route from source to destination fails because of a link outage, a soft state approach with dynamic rerouting supplied by the network will, with some probability, temporarily lose QoS. The length of time will be governed by the time required to determine a new route and process the reservation message. On the other hand, a hard state protocol, such as ATM, will always drop the connection if a route fails and require a new call setup message.

The next interesting aspect of soft state relates to group reservation support. RSVP is designed to support heterogeneity of bandwidth requirements if there are multiple receivers of a multicast session. Each receiver can get a different QoS by either merging requests or using different QoS layers. An additional benefit of RSVP is that because it is a receiver-driven protocol it has the potential to scale to a large number of participants. The large scalability is due to the ability to reduce the number of messages traveling upstream via merging. This has the benefit of possibly requiring less state in routers. Those critics who argue against RSVP believe the scalability claim is grossly overstated, since the soft state requires a steady stream of refresh messages, which, in their opinion, will not scale.

Assessing which state is optimal depends upon one's point of view. From the standpoint of complexity, the soft state approach is better if the end station never realized the outage because it was constantly transmitting RESV messages. The hard state protocol will require the end station to receive a message from the network specifying that the virtual circuit has been deleted; then the end points must reestablish the circuit.

If an RSVP reservation is successful, there is no explicit acknowledgment from the network, as there would be with an ATM-switched virtual circuit call request. This design decision is made to simplify the protocol, but it could pose problems when interworking with ATM. In cases where the reservation messages (RESV) are transmitted but then lost, the end stations may assume their request was accepted. Thus, the source may begin to transmit data to a destination that has no resources reserved for it between the two computers and will likely be dropped between the destination by the routers. In order to allow a host to determine if the RSVP message was successful, there are explicit provisions for the hosts to query the network for state information. This query response mechanism must be requested in the RSVP reservation messages and is returned with a positive acknowledgment if the request was successful.

10.2 Scoping RSVP

The preceding discussion addresses RSVP's features from a high level, but it is worth noting which functions are clearly not associated with the protocol. RSVP is not a routing protocol. It assumes the prior existence of a layer 3 routing support via protocols such as IS-IS, IGRP, BGP, and so on. RSVP only provides the ability for entities to signal their desired quality of service. It asks for state, but does not help provide it. In addition, RSVP is not an admission control or packet scheduling application. These functions, while tightly coupled with a network providing guaranteed QoS, are left to the interconnection devices. Clearly, as with ATM CAC, this is an instance where router and hub vendors can differentiate their RSVP products.

10.2.1 RSVP nomenclature

Before describing operational details of the RSVP model, relevant nomenclature used by RSVP is defined in table 10.1.

Table 10.1 RSVP Terminology

Term	Definition
Advertised Specification (ADSPEC)	This is a set of network-modifiable parameters used to describe the QoS capability (i.e., service classes supported) of the path between the source and destination.
Filter Specification	The filter specification defines the set of data packets that receive the QoS specified by the flow specifications. The session ID, an implicit part of the filter, segregates and schedules the packet classifier output according to their source address and port.
Flow Specification(FSpec)	The flow specification specifies the desired QoS reservation. The flow specification in a reservation request contains the service class and two sets of numeric parameters: TSpec and Rspec. If the request is successful, the flow specification sets the packet scheduler.
Packet Filter	The unique header pattern occurring in packet classification is known as a packet filter.
Resource Specification (RSpec)	This is the characterization of resources reserved to satisfy receivers in terms of what QoS characteristics the packet stream will use. This information evaluates QoS requests.

Table 10.1 RSVP Terminology

Term	Definition
Sender Template	The sender's IP address and, optionally, port number, are known as the sender template.
Session	A session contains the specific parameters that describe a reservation, including unique information used to differentiate the traffic flow associated with it. Each session is identified by the combination: Destination address + protocol + port.
Transmission Specification (TSpec)	This is the characterization of data flow from the standpoint of the packet stream's physical appearance (i.e., headers, packets/second, etc.). This information differentiates the QoS requests.

There are three components used by hosts to determine and signal QoS in the integrated service model. The model is relatively straightforward, because there is clear distinction of tasks. Early versions of the protocols have limited functionality. The three components are as follows:

- The setup protocol used by hosts or routers to signal QoS into the network
- A traffic model or specification, known as the Flowspec, which defines the traffic and QoS characteristics of data leaving a source
- Traffic controls or shaping mechanisms that measure traffic flows leaving a host or router to ensure that they do not exceed the agreed-upon QoS.

The interaction of these components is illustrated in figure 10.1. This figure illustrates two network devices, a host and router, involved in an RSVP exchange. The data flow is controlled so that it is conforming to the RSVP signaled parameters when metered by the scheduling process. At the ingress to the router, the flow is measured to ensure that it has not exceeded the requested bounds.

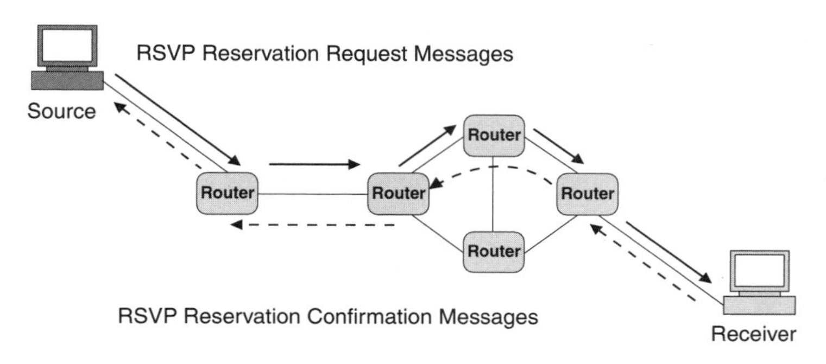

Figure 10.1 Flow specification message flow

10.3 RSVP model

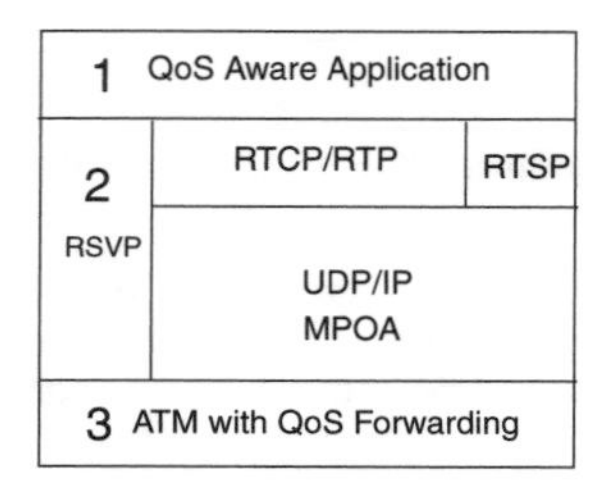

1. QoS aware applications can use the well-known RSVP interface to make generic QoS requests.
2. The RSVP signaling protocol messages are translated into a sgnaling protocol that matches the current physical layer's capabilities.
3. ATM signaling messages are generated and transmitted into the network to request the application's desired QoS.

Figure 10.2 Possible uses of RSVP and protocol stack

There are several possible uses of RSVP, since it is designed to be a very flexible protocol. One of the more likely applications in ATM networks is to use the protocol to interface directly with the application. Below the RSVP stack, there exists an interface with traffic shaping software provided by the ATM layer. This protocol stack is shown in figure 10.2.

In this model, the application developer uses the RSVP protocol as a generic means for requesting QoS. This allows software designed independently from the lower-layer protocols to be run (i.e., the application could run just as well on an IP packet-switched network). When run on an ATM network, the RSVP signaling messages translate into ATM signaling messages requesting the creation of virtual circuits in order to provide the desired QoS.

Before describing the operation of RSVP, it would be helpful to complete the discussion of the terms used with RSVP.

- The term "flow" is often used when describing RSVP, MPOA, or any of the cell-switching routers. While the definition can vary, essentially it is made up of a sequence of packets with the same QoS requirements. Typically, their IP destination address and port number segregate flows.
- RSVP uses the concept of a session to designate flows with a particular destination IP address and port. Then, a session can be isolated and provide special QoS-related treatment.

RSVP defines two terms used to describe traffic categories.

- A flow specification is the information contained in the reservation request pertaining to QoS requirements of the desired reservation
- A filter specification specifies the flows received or scheduled by the host.

These terms will be described in greater detail, along with their relation to the protocol, in the following section.

10.4 Protocol operation

The operation of RSVP centers on the exchange of RSVP messages containing information objects. Reservation messages flow downstream from the senders to notify receivers about the pending content and what would be the associated characteristics required to adequately accept the material. Reservations flow upstream towards the senders to join multicast distribution trees and/or place QoS reservations.

The information flow in RSVP can be subdivided into three categories:

1. RSVP data generated by the content source specifying the characteristics of its traffic (sender TSpec) and the associated QoS parameters (sender RSpec): This information is carried, unmodified, by interconnecting network elements in an RSVP SENDER_TSPEC object to the receiver(s). An RSVP ADSPEC is also generated by the content source and carries information describing properties of the data path, including availability of specific QoS services.
2. RSVP data generated by the interconnecting network elements (i.e., ATM switch and IP routers) is used by receivers to determine what resources are available in the network. The QoS parameters that can be reported help the receivers determine available bandwidth, link delay values, and operating parameters. As in the sender's RSVP data, an RSVP ADSPEC can be generated by the interconnecting elements; it carries a description of available QoS services. The existence of two objects, an ADSPEC and a SENDER_TSPEC, describing traffic parameters to downstream receivers can be confusing. However, there is a subtle distinction. The SENDER_TSPEC contains information that cannot be modified, while the ADSPEC's content may be updated within the network. Reasons for an update would most likely be caused by a network element setting bit patterns to signal lack of QoS support.
3. RSVP data generated by the receiver specifying the traffic characteristics from both a packet description (receiver TSpec) and a resource perspective (receiver RSpec): This information is placed into an RSVP FLOWSPEC and carried upstream to interconnecting network elements and the content source. Along the path towards the sender's routers, because of reservation merging, it is possible to modify the FLOWSPEC.

10.5 Reservation styles and flows

RSVP is a receiver-oriented protocol in which receivers send QoS requests upstream towards senders. There are several reasons the protocol designers have receivers control

QoS requests. Primarily, it is believed that the receiver, ultimately, is best informed about its local needs for quality (i.e., a local user can best determine the most desirable quality of video and if additional economic costs are justified). In addition, separating the signaling process from the source will afford better scalability. The argument involves the possibility of the source having thousands of receivers, especially if the requests to receivers and the parameters are never seen by the source.

The RSVP protocol is also designed with two reservation styles. These assist multi-sender sessions and select a subset of senders identical to the IP multicast model. ATM only supplies point-to-multipoint connections with a single sender. The reservation styles supported are as follows:

- Distinct reservations, which require separate reservations for each sender
- Shared reservations, which can be shared by multiple senders and also allow receivers to dynamically change the list of senders (i.e., change channels)

With the above definitions in mind, we can begin to describe the RSVP information flows. The RSVP flows and related terminology are illustrated in figure 10.3.

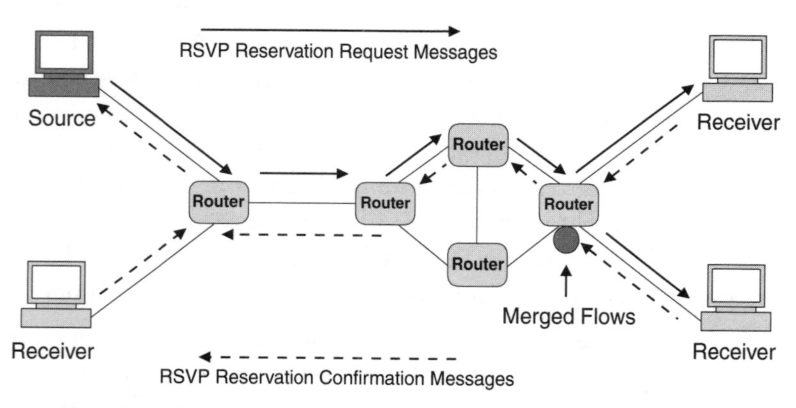

Figure 10.3 RSVP flows

10.5.1 Reservation styles

Resource needs vary widely depending upon the application—for example, audio and video conferencing may require a very high likelihood for the audio section of the content to provide an acceptable quality for the listener. However, data transmission more important than the Internet's default best-effort service may require a higher quality of service; it may only need to specify a few parameters and delay may not matter. An example of this would be a virtual private network run over an ISP's backbone network. In order for a corporation to rely upon an ISP for intracorporate communication, it must know there is a low probability of loss due to congestion. The ISP can then obtain a larger return on its infrastructure investment by selling its backbone at both a regular rate and a premium rate for customers using RSVP.

In the RSVP model, applications have the ability to request different reservation styles, depending upon the type of service requested or economic considerations. Currently, there are three reservation styles supported. They are shown in table 10.2.

Table 10.2 Reservation Styles

Reservation Style	Definition
Fixed Filter	One reservation per source
Shared Explicit	One reservation shared by listed sources
Wild-card Filter	One reservation for all sources

The wild-card filter reservation style is designed for sessions in which all sources require similar service guarantees and the sources are able to limit their output, as in an audio conference. In a wild-card filter, one reservation is made and shared for all sources. This style would be particularly useful in the case of a multicast session.

The fixed filter and shared explicit reservation styles differ from a wild-card filter, because they are designed for cases where several different reservations must coexist. The fixed filter reservation uses only one reservation per source and supports applications, such as video distribution, that need one data stream per source. With the shared explicit reservation, one reservation is shared by several sources. This type of reservation would be useful if multiple users were concurrently viewing two different video sources, each of which required a different quality of service.

The one problem remaining with reservations involves the difficulty of merging multiple requests for the same source when the source is multicasting to the receivers. The difficulty occurs if each receiver has specified different parameters when asking for dedicated bandwidth—for example, each receiver may pick values it believes are acceptable for the TSpec and RSpec. However, because these are multidimensional, the requested state for the same source can be very different.

The ideal solution would be to provide a means for merging multiple heterogeneous requests by somehow picking the "largest" request. With this goal in mind, there are two distinct problems that should be avoided. First, if the merge is too conservative, requests may result in over-reservation. Second, if the largest values are gleaned from the

TSpec and RSpec without care, then the superset could generate a request that is too large to be accepted, whereas the individual requests would have succeeded.

10.5.2 Multiple multicast groups

To help solve this problem, the concept of Multiple Multicast Groups (MMG) was developed. With MMGs, the source is responsible for defining the division of the total traffic into layers of varying qualities. When receivers attempt to join a session, they are placed into a homogeneous group based on their local capabilities. The receiver's TSpec and RSpec may not be identical to the group's parameters, but the likelihood of success is much higher.

There are two possibilities for addressing the problem solved by MMGs. First, the network could use multiple data flows within the same RSVP session but assign different ports from the same source for the different streams. This approach would work; however, there would be some waste of resources, because if heterogeneity of receivers exists, then not all packets go to the same set of receivers.

MMGs provide a method to successfully differentiate data flows in a way that works with the multicast routing protocols by using different multicast addresses. If multiple substreams were encapsulated in one session, then, in all likelihood, some of the substreams would be dropped prior to some datagram receivers downstream requesting the flow. After submitting a request, the receiver can query the network to determine what state is reserved for the session it has joined.

In developing the RSVP model, the IETF designed a description of the traffic pattern contained in the reservation request sent by the host or router. This description is called the TSpec and is specified by a set of parameters that are very similar to ATM's traffic models [1]. The TSpec is communicated to the network to specify the rate of data transmission and is then used as a traffic shaping descriptor. The TSpec is defined as a *token bucket* with a value for the bucket size and the data transmission rate. Data sent by a host or router cannot exceed the value of $rT+b$, where T is a time interval, r is the arrival rate, and b is the bucket depth. The concept of the TSpec is illustrated in figure 10.4.

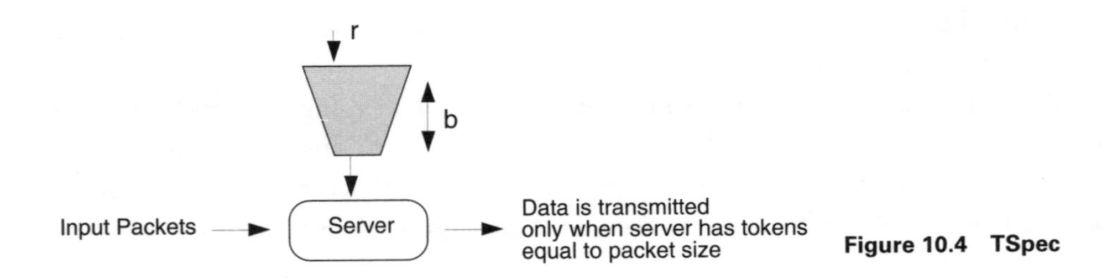

Figure 10.4 TSpec

10.6 RSVP messaging

Implementations of the RSVP protocol are very similar to client/server models. The higher-level protocol description dictates message types that are exchanged and specifies which sequences are supported. The RSVP protocol also defines several data objects, which carry resource reservation information but are not critical to RSVP itself.

There are five basic message types used in RSVP and each message type carries several sub-fields, as shown in table 10.3.

Each of these message types is used at a particular interval in the establishment of the RSVP state (see figure 10.5).

Table 10.3 RSVP Message Types

Message Types	Funciton
CONFIRMATION	Sent by a receiver, this optional message signals successful resource reservation
ERROR	Notifies an abnormal condition such as a reservation failure
PATH	Sent by the source to specify that a resource exists and, optionally, which parameters should be used when transmitting
RESV	Transmission of a message in hopes of reserving resources
TEARDOWN	Deletes an existing reservation

10.6.1 PATH messages

The protocol operates by the source sending a quasi-periodic PATH message, out of band from the actual reserved quality data session, to the destination address (i.e., receivers) along the physical path that joins the computers*. As the PATH datagrams traverse the network, the interconnecting routers consult their normal routing tables to decide where to forward the message.

When a router processes the PATH message, it will establish some "PATH state" gleamed from fields in the message. PATH state records information about the IP address of the sender along with its policy and QoS class descriptions.

Upon reception of the PATH message, the receiver will determine that a connection has been requested and attempt to determine if, and how, it would like to join the session. The receiver will not join a session by using the PATH sender's IP address, but

* In the case of multicast, receivers must join the multicast group before they can receive PATH messages.

0		1	2	3
Version	Rags	Message Type	RSVP Checksum	
Send_TTL		(Reserved)	RSVP Length	
Object Length			Class Number	C-Type
Object Contents				

Version specifies the protocol's version number
Flags field is undefined at this time
Message Type represents the RSVP signal
1 - Path
2 - Resv
3 - Path Err
4 - ResvErr
5 - PathTear
6 - ResvTear
7 - ResvConf
Checksum is calculated over the message
Send_TTL cooresponds to the IP TTL of message including variable length headers
RSVP Length signifies total length of message including variable length headers
Object Length indicates length of the individual variable length objects
Class Number specifies the object clas
C-Type represents the object's clas type

Figure 10.5 RSVP packet header

will use the address specified in the SENDER_TSPEC, because the source could be a Class D multicast address.

PATH messages contain the following fields:

- Session ID
- Previous hop address of the upstream RSVP neighbor
- Sender descriptor (filter + TSpec)
- Options (integrity object, policy data, ADspec)

The PATH messages are sent at a quasi-periodic rate to protect the systems from changes in state. If a network failure causes the route the PATH messages took to change, then the next PATH will reserve resources on the next cycle. If there are interconnecting devices along the old path unable to be reached, their stored state will timeout when they do not receive the quasi-periodic PATH message. The PATH message contains the previous hop address of the upstream RSVP neighbor. The previous hop address is used to ensure that the PATH message has traversed the network without looping. Finally, the PATH message contains a SENDER_TEMPLATE object, which is simply the sender's IP address and is used for identification.

10.6.2 RESV messages

If the receiver elects to communicate with the sender, it then sends a reservation message upstream along the same route the PATH message used. If the RESV message fails at one of the intermediate routers, an error message is generated and transmitted to the requester. In order to improve network efficiency, if two or more RESV messages for the same source pass through a common router or switch, the device can attempt to merge the reservation. The merged reservation is then forwarded as an aggregate request to the next upstream node. The RESV messages are addressed to the upstream node with the source address becoming the receiver. The RESV contains a TSpec corresponding to the session's source.

RESV messages contain the following fields:

- Session ID
- Previous hop address (downstream RSVP neighbor)
- Reservation style
- Flow descriptor (different combinations of flow and flow specification are used, based on reservation style)
- Option (integrity, policy data).

If the request is admitted, then, in addition to forwarding the RESV messages upstream, the host or router will install packet filtering into its forwarding database. The forwarding database is queried when the device has a packet to be transmitted and is used to segregate traffic into different classes. The flow parameters established for this QoS-enabled traffic will also be passed to the packet scheduler. The scheduler forwards packets at a rate compliant to the flow's description parameters.

If the interconnecting network contains routers that do not understand the RSVP protocol, the PATH/RESV messages are forwarded through the non-RSVP cloud, since they are just regular IP packets. As shown in figure 10.6, the routers at the edge of the RSVP system communicate with their neighbor as if they were side by side. The protocol will operate in this environment; however, the quality of the reservations will be mitigated by the fact that the network now contains spots providing best-effort performance. The performance across these spots must be estimated and communicated to the receivers in ADSPEC messages.

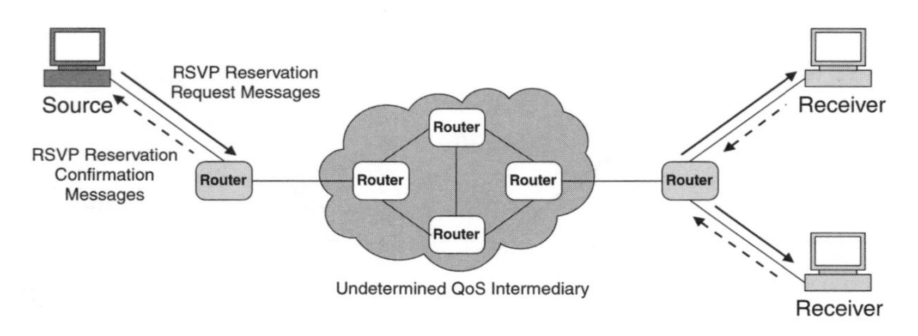

Figure 10.6 Non-RSVP intermediate networks

The above messages perform the heart of the RSVP protocol. The following miscellaneous message types add features not critical for operation. In fact, because RSVP is soft state, the protocol could function without these additional messages.

10.6.3 TEARDOWN message

An additional message in the specification is known as the TEARDOWN. It is issued when either side of a session wishes to terminate communication. In this event, the device generates a TEARDOWN message which propagates across the network and releases the reservation, thus freeing resources for other users.

There are two types of TEARDOWN messages: PathTear and ResvTear. If the source issues the termination request, then a PathTear is generated. In the case of multicast, all resources associated with the session will be deleted as the message propagates down the multicast tree towards the receivers. This message will remove the source's state, in addition to all reservations made for the source's data flow.

When a receiver initiates a termination, it does so by sending a ResvTear message. This message travels upstream and differs from the PathTear, since it only removes the reservation state of the receiver. The message will continue up a multicast distribution until it reaches a merge point where one leg is still active; then it stops.

The TEARDOWN is optional with RSVP, because the protocol is soft state, all reservations will automatically expire if they are not manually refreshed. The only drawback to reliance on the timeouts for reclaiming resources occurs when the time is set high. In this case, the resources will be waiting for their timer to expire and even though they are no longer used, the resources will be locked out from other sources.

10.6.4 ResvConf message

RSVP allows a receiver to query the network to determine what state has been reserved for its session. The explicit RESV_CONFIRM object is required in a RESV message because the protocol's *single-pass* model [2] does not provide the receiver with any information about the successfulness of its request. A RESV message, with a RESV_CONFIRM object containing the IP address of the receiver is passed upstream towards the source and, at each router, the flow specification associated with the reservation is compared with reserved reservations from the downstream router. If the reservation in the current router is equal to, or larger than the receiver's, the RESV_CONFIRM forwarding stops and the receiver is returned a ResvConf message with Flow specification values from the current router.

There are two obvious problems associated with this approach. First, the messages are transmitted unreliably and may be lost. However, all RSVP messages support this feature, so the user should be aware of that fact. The second problem occurs with cases involving multicast merging. Before the messages reach the source, they may encounter a router with a reserved state from a previous RESV of another receiver. Consequently, the merged RESV fails at some point farther upstream. In this case, the router sends a ResvConf downstream, signifying that the RESV has been successful when, in fact, it has failed.

10.7 Operational procedures

An application wishing to make use of RSVP signaling communicates with the protocol (i.e., creates the appropriate signaling messages) through an Application Program Interface (API). Before receivers can make reservations, the network must have a clear understanding of the source characteristics. This information is communicated across the API when the hosts register themselves. The RSVP code in the host then generates a SENDER_TSPEC object which contains details of the resources required and what the packet headers will look like. The source also constructs the initial ADSPEC containing generic parameters. Both of these objects are then transmitted in the PATH message.

As the PATH message travels from the source to the receivers, routers along the physical connection will modify the ADSPEC to reflect their current state. The traffic control module in the router will check the services requested in the original ADSPEC and the parameters associated with those services. If the values cannot be supported, the ADSPEC will be modified, or, if the service is unavailable, a flag will be set in the ADSPEC to notify the receiver. By flagging exceptions, the ADSPEC will notify the receiver of the following:

- There are non-RSVP routers along the path (i.e., links that will provide only best-effort service)

- There are routers along the path that do not support one of the service categories, controlled-load or guaranteed,
- A value for one of the service categories is different from what is selected in the SENDER_TSPEC.

At the receiver, the ADSPEC and SENDER_TSPEC are removed from the PATH message and delivered to the receiving application. The receiver uses the ADSPEC/SENDER_TSPEC combination to then determine what resources it needs to receive the contents from the network. Since the receiver has the best understanding of how it interacts with the source application, it can accurately determine the packet headers and traffic parameter values for both directions of the session from the ADSPEC and SENDER_TSPEC. Finally, the receiver's Maximum Transfer Unit (MTU)* must be calculated, because both guaranteed and controlled-load QoS control services place an upper bound on packet size. Each source places the desired MTU in the SENDER_TSPEC, and routers may optionally modify the ADSPEC's MTU field on a per-class-of-service basis.

Once the receiver has identified the parameters required for the reservation, it will pass those values to the network via its RSVP API. The parameters from the TSpec and RSpec objects are used to form the flow specification, which is placed in an RESV message and transmitted upstream using the default route. When an internetwork device receives it, the RESV message and its corresponding PATH message are used to select the correct resources to be reserved for the session.

10.8 Interworking RSVP with ATM

In this section, we will examine the differences between RSVP and ATM signaling models and describe proposals that address how these dissimilar technologies will be internetworked. As described in the previous chapter, there is a compelling need for integrated real-time service support that includes IP routing, ATM-style quality of service, and multicast features. These services should be available to applications running on ATM networks, IP networks, IP over ATM networks, and mixed networks. In this model, there are several possible communication paths between hosts. Some of the possibilities are as follows:

- ATM host to ATM host
- IP host to IP host

* The MTU is the maximum packet size that can be transmitted. An MTU is specified to help bound delay.

- ATM host to ATM host over a non-ATM IP network
- IP host to IP host over an ATM network
- ATM host to IP host.

To enable an integrated IP/ATM network, the features of the ATM model, the IP model, and quality of service must be interconnected within the network and available at the application level. The goal of the integrated services model is to mask the underlying technology from the application while still providing the following features:

- Internetwork routing, which allows applications to achieve their desired performance from the network via optimal path selection.
- Multicast capability—which permits one-to-many, or many-to-many communication flows.
- Quality-of-Service facilities per which are parameters describing the desired characteristics applications can expect from the network

When an IP datagram, which is QoS aware, is passed from a packet switched network interface to an ATM cloud, it must cause some type of connection setup message to occur in the ATM network. The difficulty involved in this is the determining ATM traffic class appropriate for the IP data stream, and, once the correct traffic class has been selected, what parameters should be used in the ATM signaling message. The following is a list of some of the potential problems related to integrating ATM and RSVP.

- ATM's resource reservation is sender-based versus RSVP's, which is receiver-based.
- The sender controls ATM's point-to-multipoint connection management.
- ATM maintains a *hard*, unchanging connection state. RSVP maintains a *soft*, changeable state that must be continually refreshed.
- ATM QoS resources are set up at connection time; RSVP QoS is independent of connection setup.
- ATM has a single QoS specification for point-to-multipoint sessions. RSVP supports heterogeneity of QoS specifications for receivers.

When comparing the IETF's and the ATM Forum's traffic and QoS descriptions, the reader may notice some differences, especially with the traffic descriptors. The IP traffic description contained in the TSpec uses the same parameters that correspond to VBR's SCR and MBS. Therefore, the TSpec uses the same parameters as the values of the second Leaky Bucket implemented in ATM's UPC. As with ATM's peak cell rate, the TSpec also supports a peak rate parameter.

The service descriptor, RSpec, is somewhat different between RSVP and ATM. In RSVP the RSpec specifies distinct categories, which indicate relative quality levels. On the other hand, ATM's cell transfer delay could be used to indicate different levels but currently is a discrete value. One point to consider with the controlled delay service is that the end routers (i.e., routers at the edges of an ATM network) could perform measurements on the ATM network and allocate a portion of the permitted total delay to the ATM cloud.

The values of ATM's cell loss parameter pose a somewhat larger problem, because there are no corresponding values in IP services. In order to internetwork RSVP signaling to ATM signaling, the edge router must choose an appropriate cell loss ratio, based on average or worst-case values, for the ATM virtual circuit.

An additional area of difference is the mapping of service classes. The IETF's model contains three service classes, whereas the ATM Forum has defined five categories. Therefore, there are several possible mappings that can be supported when converting RSVP to ATM signaling, or vice versa. To a large degree, the problem of mapping different signaling protocols to each other is essentially a matter of network provider policy or end-user requirements—that is, a service provider will provide a limited subset of the total possible set of ATM signaling messages. This will be done to reduce the complexity of end-user education and carrier billing/monitoring systems. Some of the possible mappings are as follows.

- The guaranteed service's most likely fit is the ATM constant bit rate category. The values of PCR, SCR, and MBR can be mapped directly from the TSpec to the RSVP reservation request receiver TSpec; the only "guess" needs to be the value of the CTD.
- The controlled delay service has explicitly specified values of delay that can be mapped to ATM's real-time VBR service category. It would be possible to utilize ATM's CBR in this case as well; however, the CBR category does not yield much statistically multiplexing gain, so it would be suboptimal for Internet traffic.
- The final category, best effort, can be mapped to any of ATM's non-real-time services, such as UBR or ABR, or non-real-time VBR. The most common choice today is UBR in concert with early packet discard.

One of the more profound differences between the two protocol models is RSVP's reservation style and ATM's signaling style. In ATM's model, QoS parameters are specified at connection setup by the sender and remain constant throughout the lifetime of the virtual circuit. In the RSVP model, senders advertise content and traffic parameters towards their receivers. It is then the receiver's responsibility to make the reservation request. Interworking these different systems is cumbersome, though possible. One technique is to have senders transmit PATH messages using virtual circuits created with ATM's UBR service class. When a receiver decides to reserve resources for a session, it can then use the parameters in the PATH message to create a bidirectional virtual circuit with the sender.

Another difference is due to RSVP's soft state resource reservations and the problems associated with dynamic resource negotiation. In the RSVP model, resource reservations must be refreshed periodically or they will time-out and be deleted. In contrast, ATM resources are allocated at connection setup and remain intact for the duration of the connection. They are removed only when the network receives an explicit connection teardown message. RSVP therefore allows the requests to be dynamically modified over time. ATM, on the other hand, requires a new virtual circuit for QoS modifications.

The issues of heterogeneity among receivers are somewhat complex and clearly important for the Internet services. Heterogeneity allows many different receiver types to concurrently access the same source; for example, a video source could concurrently be viewed by T1, ISDN, and V.35 modem Internet connections. In order to support heterogeneity among receivers when interworking with ATM, and not have a unique virtual circuit to every destination, the source must be able to *layer* its traffic.

Layers can either be a subsection of the total content (i.e., every other frame in the case of video) or a subsection of the data (i.e., just the voice portion of a multimedia session). Each receiver receives a number of layers corresponding to its local capability. Using different virtual circuits for each layer could accommodate the ATM interworking with a layer system. The problem is then reduced to management of the multiple multicast sessions.

Finally, the last area of contrast to ATM's semantics with RSVP involves internetworking with routing protocols. The RSVP model is designed to be independent of network components such as routing, and only assumes the existence of a path between source and destination. QoS negotiations occur after the data path is established, and, because some services calculate delay, RSVP relies upon the PATH and RESV messages following the same physical connections.

On the other hand, ATM resource reservation is governed by the cooperation of the signaling protocol with the PNNI (interswitch routing) protocol. With PNNI, the desired resources are examined at call setup time and then the best path is selected based on what is best for that call. There is no guarantee that the PATH message and the RESV message will follow the same physical link. This problem has no clear solution in ATM and, until one is found, developers must hope that either the physical path will be the same for RSVP messages or the QoS values will be similar, regardless of which path is selected.

10.9 Summary and outstanding issues

This chapter has presented an introduction to the integrated services model-signaling protocol, RSVP. This protocol has received a great deal of attention and promises to be a valuable tool for real-time applications. Wide-scale deployment has yet to be realized; however, if RSVP proves successful, it can, in part, be attributed to design criteria requiring robustness and scalability. While developing the protocol, a great deal of emphasis was placed on experiences gained with multicast protocols on the Internet. That experience may yield techniques that scale to size, thus keeping pace with the Internet's growth.

When interworking the IETF's and the ATM Forum's models for QoS negotiation, it is clear that many aspects of the protocols are very similar. Several of RSVP's parameters map directly to ATM's traffic parameters, and that fact may dramatically increase the speed in which the two services are deployed in tandem. The areas in which the protocol overlays are "gray" are essentially a matter of a network provider policy, which can be addressed via policy direction statements or signaling policy restrictions. RSVP policy will play a critical role in making the protocol successful. If the protocol is used without a policy, then end users may have to abuse RSVP's abilities in order to get better perform-ance for non-mission-critical applications. With an intelligent policy implementation, ISPs will be able to segregate RSVP traffic into a category that generates more revenue and corporations will be able to subdivide their traffic into multiple, monitored priorities.

From the discussion in the preceding chapters it should be clear that ATM's connection semantics are generic enough to accommodate any situation—the real difficulty is implementation once the policy has been set. With this in mind, it would be helpful before ending this chapter to present a brief review of some of the implementation difficulties and outstanding questions. To begin with, there are several points of which we have little understanding that no doubt will remain with us until large product integrated services networks are in place:

- The kind of performance one can expect with RSVP
- The specifics (latency, jitter, etc.) of establishing a flow
- The number of flows capable of being established
- How the number of flows affects the performance of establishing new flows or tearing down old flows
- When/if the problem of bundling or flow aggregation is to be addressed and how aggregation will affect the performance parameters.

Next, there is the issue of cost involved with the RSVP reservation method and refresh scheme. Until further working knowledge is gained there is concern that refreshes which support thousands of sessions between two RSVP routers could generate a substantial amount of traffic. This problem is related to the soft state protocol and only seems to be mitigated if RSVP "hardens" soft states.

The issue of cost involves the requirement to maintain the soft state information. RSVP requires that both PATH and RESV state be maintained for each flow. Current estimates for the amount of memory used in the router for each reservation are between 500 and 600 bytes. This means that approximately 1 MB of memory can maintain state for 2,000 flows. While this may seem like a large number, some estimates place the need for high QoS at millions of concurrent flows, which may be very expensive to support in terms of router memory. One last point on this topic is that this problem is not RSVP specific. It may be true for any connection-oriented network, such as ATM, that tries to fill the shoes of IP on the Internet.

Additional issues occur with the integrated services protocol suite, which, as specified, has some potential shortfalls due to its use of UDP/TCP port numbers in the filter specification.

- Because the IPv6 protocol allows users to insert variable headers, the difficulty of packet classification can dramatically change on a packet-by-packet basis.
- Several security schemes in use today implement packet encryption, which effectively deletes port numbers; therefore, the intermediate router's packet classification will not work, restricting RSVP's usefulness.

When considering the interworking with ATM, little discussion has focused on its implications for multicast, and a few illustrative points would be interesting. When running IP multicast over ATM, there are two proposed ways of implementing the internetworking: full meshes of virtual circuits between the source/destination pair, and multicast servers.

Both of these solutions will work for small logical subnetworks; however, there are some serious scalability issues to be resolved. First, full meshes can exceed the amount of virtual circuits supported in an ATM interface. At the time of writing, that number is about 2,000 virtual circuits. The second scaling problem is with multicast servers, because they can become the bottleneck for multicast replication/distribution.

Even with these scaling problems, these solutions are specified in the standards and will be implemented. Therefore, network designers must understand these limits and apply the multicast technology that makes sense for a particular application. Trying to take one approach and applying it to all multicast scenarios simply will not work, due to resource limitations and/or performance barriers.

Finally, there is the very real issue that RSVP is designed to pass traffic through a congested network when the best-effort packets have filled links to capacity. When large numbers of users migrate to RSVP and the RSVP bandwidth is then used up, the real-time applications block the non real-time applications, putting us back in the same situation that exists today. This problem will force the issue of continued research into policies for control and measurement purposes. Unfortunately, it may also lead down the difficult path of telecommunication-style billing systems imposed on Internet-style providers.

Although solutions to many of the resource reservation and interworking problems have been developed, the major challenge now is to develop implementations that will efficiently determine and signal, at run time, QoS. The differences between the RSVP and ATM models of operation are significant but not insurmountable.

10.10 References

1 ATM Forum, *ATM User Network Interface Specification Version 4.0*, Englewood Cliffs, NJ: Prentice Hall.

2 S. Shenker. and L. Breslau, *Two Issues in Reservation Establishment Paper,* ACM SIGCOMM '95, Cambridge, MA, August 1995.

PART 4

Migrating to Multiprotocol Networking and Alternative Technologies

CHAPTER 11

MPOA LAN Communication: ATMF and IETF Approaches

Up to this point, our discussion has addressed signaling, models, and transport protocols that permitted applications to request different Classes of Service (CoS). In this chapter, we will shift the focus away from high-level protocols and towards the application of these protocols in MPOA networks. This chapter will take a phased approach in addressing the techniques used to run QoS-enabled applications over ATM.

First, we will discuss protocols that are being used today to help utilize ATM as the physical communication medium in a local area network without necessarily utilizing QoS features. This information is critical for network managers and designers, because the technology described in this chapter provides the foundation of MPOA LANs—that is, computers connected to MPOA networks will use the protocols described in this chapter to communicate with their LAN neighbors. After our discussion of LAN protocols, we will diverge into a discussion of technologies that expose ATM's QoS and how these technologies are used in network design.

11.1 Scope of local area ATM networking problem

There are several problems network managers face when migrating their legacy networks to ATM, regardless of whether the transition is to a simple network or a complex MPOA system. Several of these problems are associated with joining a connection-oriented ATM network with a connectionless LAN similar to Ethernet technologies. When joining these different technologies, the problems encountered can be subdivided into two categories:

- Address resolution—final determination of where the destination resides
- QoS requesting and mapping—configuration and creation of a communication path between the source and destination.

The ATM Forum and the IETF have devised several approaches for deploying ATM networks that coexist well with legacy LANs. When selecting a migration path from existing LAN infrastructures to QoS-capable, ATM-based internetworks, network managers need to consider the strengths and weaknesses of each approach. Fortunately, as will be described, there are several areas of overlay where the approaches of the ATM Forum and the IETF are almost identical.

To best illustrate the components used when migrating legacy networks to ATM, this information is divided into sections that address standards developed by different bodies. The first covers the details of both the ATM Forum's LAN Emulation (LANE) [1], followed by the IETF's Classical IP [2]. These protocols make up the most common elements used when migrating from legacy LAN environments. Building on these two sections, we will discuss the topics of shortcut path selection, flow modeling [3], and distributed routing, which essentially form the foundation of the ATM Forum's Multiprotocol over ATM (MPOA) work [4]. The concluding material then addresses alternative techniques for building ATM networks with the Cell Switching Router (CSR) [5], in addition to other issues relevant when migrating to an ATM framework.

11.2 LAN Emulation background

Prior to LANE, early implementers of ATM encountered significant difficulty due to the connection-oriented nature of ATM. This occurred because most network protocols, such as Ethernet or Token Ring, were designed as connectionless networks. On connectionless networks, any host can communicate with any other host simply by placing a packet onto the network. The packet is then transmitted to all hosts attached to the local network without any further intervention by the originating host.

If the packet is destined for a computer that is a member of the local network, then the packet will be seen by that machine and read directly from the LAN. If the packet is destined for a host outside the local network, then it is the responsibility of the local router to forward the data to the correct destination. If the ultimate destination is several networks away, the process is repeated until the packet reaches the network where the destination computer resides. This type of communication—announcing the data packets to all the hosts and forwarding through routers—is possible on connectionless networks because all the hosts share the same physical medium.

On the other hand, connection-oriented networks are much like telephony, in that they require an established connection prior to communication. As can be seen from the discussion of connectionless networks, this is completely opposite the manner in which legacy LANs operate. There are additional complexities associated with connection-oriented networks, due to their difficulty with broadcast messages. Broadcasts are a special message type and, once transmitted by the source, will be relayed to all destinations on the local subnetwork. The ability to transmit a message to all hosts on the network via broadcasts can be cumbersome with ATM because it requires either a unique virtual circuit to all destinations or some type of multicast server. (See figure 11.1)

In addition, since virtual circuits provide for one-to-one communication, a host broadcasting to a set of other hosts, as would be the case in a multicast video conference, must be aware of any additions or deletions to the receiving set—that is, when any new device joins or leaves the multicast group, changes in virtual circuits must be made. This problem has been a difficult one to solve with connection-oriented networks such as ATM and will continue to be an area of active research for some time.

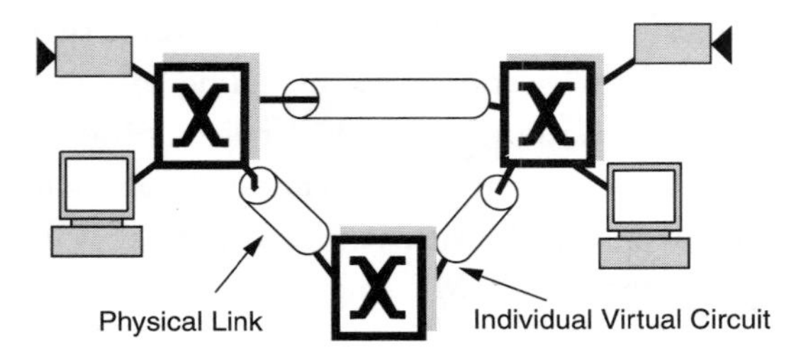

Figure 11.1 Connection-oriented cell switching

Though a multicast server can be very helpful in solving the problems associated with a small network, there are serious scalability problems that occur when a single device becomes the focal point in the network. The issue of running multicast applications will be covered in greater detail later in this chapter, but for now the important point to consider is that the lack of a broadcast (connectionless) capability is the major impediment network managers face regarding running existing applications over ATM easily. In fact, the only applications running on ATM without any fundamental software modifications that is, adding LANE or Classical IP support—are those designed from scratch to implement ATM signaling for virtual circuit creation.

The constraints imposed by ATM's connection-oriented methodology, coupled with a lack of destination address-resolving SVC software, such as LANE, resulted in the first ATM networks using only PVCs. However, as these networks grew, manually establishing PVCs for each pair of communicating hosts became cumbersome. While this could be tolerated for networks with dozens of interconnected computers or routers, it quickly became problematic in a network of hundreds of hosts, where any host could communicate with any other host.

Lack of switched virtual circuits on the ATM switches is offset by statically configuring the network, which means manually instructing routers and computers on how to reach each other across the ATM network. This proved effective for prototype systems, but was difficult to use and not scalable as ATM became a core enterprise technology. To improve the ability to "plug and play" and increase scalability, the ATM Forum formed the Local Area Network Emulation Over ATM working group, which was chartered to develop protocols that allowed quick and easy use of ATM. The requirements set forth by the ATM Forum LANE group were as follows:

1. Should be based on the user network interface specification version 3.0
2. Provide both high performance and scalability
3. Capable of protocol independent switching across logical LANs
4. Capable of seamless interworking with legacy LANs via bridges
5. Capable of supporting PVCs, SVCs, or any combination

In an effort to achieve the above goals and facilitate the deployment of ATM technology, the ATM Forum has been very aggressive in developing techniques for interconnecting installed legacy networks to new ATM systems. In particular, requirement 3 has been met, because it is very easy to build networks where legacy LAN-attached hosts have no idea that they are communicating with ATM-attached hosts. Furthermore, applications running on ATM hosts are also unaware they are using ATM as a network medium. The method used to reach this goal is the LAN Emulation (LANE) specification. LANE's name is very appropriate, because when the protocol is implemented, the behavior of legacy LAN protocols such as Ethernet is emulated on ATM.

By emulating the behavior of legacy networks, LANE's goal is to provide support to ATM users faced with the problem of interconnecting their installed base of LAN protocols over a new ATM medium, while at the same time minimizing the impact to their existing systems. It should be noted that the techniques used by LANE to solve the generic problem of computers communicating are reused in MPOA networks.

LANE's features can be summarized as follows.

- LAN Emulation provides a mechanism for existing LAN-based client/server applications to run over ATM networks without modification.
- LAN Emulation uses ATM as a backbone to interconnect existing legacy LANs to achieve higher bandwidth.
- LAN Emulation permits several emulated LANs or Virtual LANs (VLANs) to concurrently share the same ATM network. This allows one physical network to appear as several logical networks.
- LAN Emulation may be deployable in ATM networks immediately.

Though currently very functional, the LANE protocol suite is evolving, unfinished work, which will continue to be developed and enhanced for years to come. While the ATM Forum has produced a workable specification, additional work is needed, particularly in the areas of redundancy and scalability. Over the course of the next few years, subsequent versions of the LANE specification will be produced and modifications may be made to existing documents to realize the LANE version 2 specification. The work on LANE version 2 is being shared by the original LANE working group and the MPOA group within the ATM Forum.

One of the chief benefits of LANE is the ability of all devices attached to a LAN Emulation network to function in a plug-and-play fashion, requiring minimal configuration. This ease of use is primarily due to the fact that application programs use the network services via standard device driver Application Program Interfaces (APIs). By emulating the API of a standard Ethernet network interface to the higher-layer applications, the LANE software can allow these applications to run unmodified on ATM networks.

There are several common application network interfaces in use today and two of the major ones are NDSI and ODI [6]. In order to effectively emulate OSI data link layer protocols, LANE offers the same services via the LANE interface to upper protocol layers. Consequently, when a network manager wants to use LANE on a computer, he or she must augment the NDI/ODI device driver with LANE emulation software and install an ATM interface in the computer. The client then loads the device driver and uses that code for the layer protocols between applications and the ATM network interface card.

The LAN Emulation protocol stack provides all of the functions commonly used in device driver APIs—NDIS and ODI [7]—and the problem of hiding ATM from applications is solved by replacing the NDIS device driver with LAN Emulation. The new device driver hides ATM by recreating the traditional API used between applications and Ethernet or Token Ring device drivers. (See figure 11.2.)

Emulation of Ethernet and Token Ring is very attractive if users are interested in easily migrating their production networks to ATM, because the two media types make up the vast majority of all enterprise LAN networks. However, it is important to realize the implications, both good and bad, of hiding the ATM network from clients and servers with LAN Emulation.

Clearly one of the major highlights of LANE is that by hiding ATM, hosts are not required to understand all the potential complexities of operating on a connection-oriented network supporting multiple levels of QoS. On the other hand, hiding ATM does

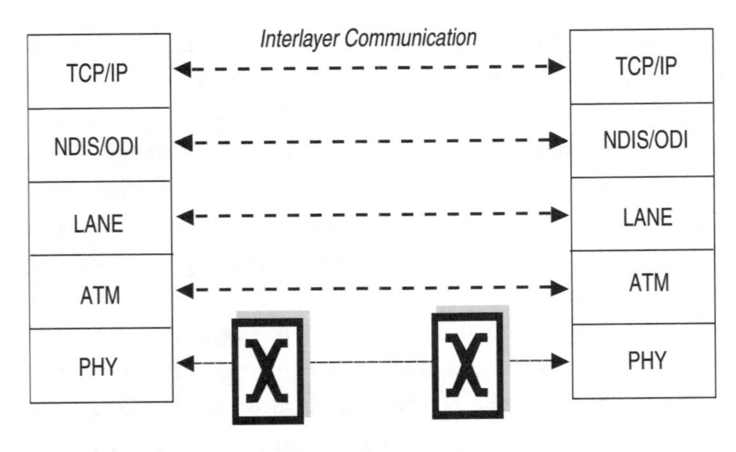

Figure 11.2 LAN Emulation protocol stack

not allow applications to utilize the technology to its fullest or make use of QoS, because there are no provisions made for communicating quality-of-service requests. This is, ironically, one of the fundamental strengths that the founders of ATM envisioned.

LANE also suffers from the drawback that, by definition, it behaves like protocol-independent bridging. Years of network design experience have shown that bridging, while effective for interconnecting small workgroups, does not scale well to support large networks. This is due to the fact that as networks grow, they are prone to experience a large load of broadcast messages when hosts try to resolve MAC addresses into layer 3 IP address mappings. The broadcasts are passed through the bridges and can begin creating a substantial load, unless some sophisticated measures are taken to reduce the traffic. If the networks were constructed from routers, then the broadcasts would not cross their subnetwork's router boundaries. Therefore, because LANE is emulating a legacy LAN, the total number of hosts attached will be limited by two factors:

- The speed of LANE processes running on end stations and the LANE servers
- The traditional rules used to measure the total number of hosts that should share a single LAN.

It should also be noted that LANE only supports emulation of one type of network at a time. If a host on an emulated Ethernet wants to communicate with a host on an emulated Token Ring, the packets must pass through a router that is a member of both emulated LANs. The positive side to LANE's bridging is that it will, unlike the IETF's Classical IP protocol, support any layer 3 protocol designed to operate on a connectionless network (e.g., IPX, IP, AppleTalk, etc.).

11.3 LAN Emulation components

To overcome the lack of connectionless capability, LANE provides the functions of an Address Resolution Protocol (ARP) via three servers. ARP is a process whereby a computer determines the mapping of two independent addressing schemes. Typically, this process is associated with mapping an Ethernet address to an IP address. Because LANE is responsible for emulating the Ethernet, it must also provide a means for ARP emulation. The servers that perform the functions of LAN Emulation and address resolution are as follows. (See figure 11.3.)

- A single LAN Emulation Configuration Server (LECS): It provides configuration information.
- LAN Emulation Server (LES): It implements address registration/resolution.

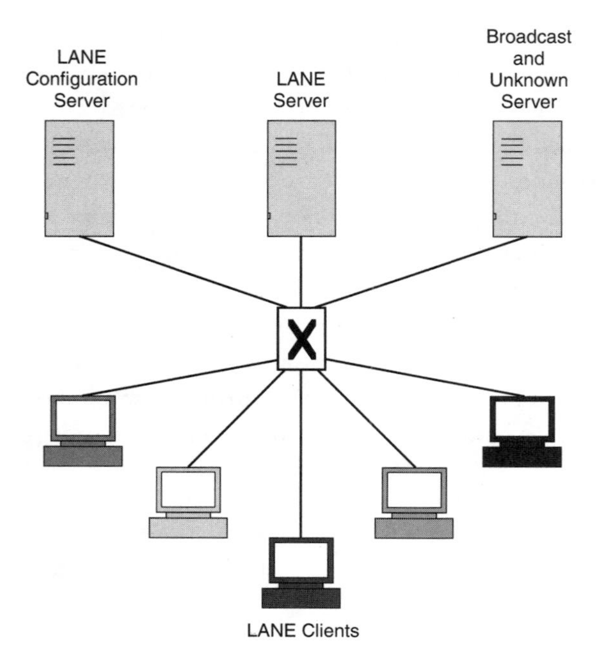

Figure 11.3 LAN Emulation Servers

- Broadcast/Unknown Server (BUS): It performs all broadcasting and multicasting functions.

11.3.1 LAN Emulation Configuration Server

The LAN Emulation Configuration Server (LECS) acts as a central point for maintaining configuration information in the ATM network. The LECS is the first point of contact when new clients boot and need to register with their ATM network prior to communication with adjacent hosts. The LECS is responsible for keeping a database of all clients within the administrative boundaries (e.g., marketing and engineering) and knows to which emulated LAN the host belongs. This information is known because, at the conclusion of the configuration process, the LECS will point the clients to their subnetwork's own ARP server.

The LECS can connect clients to different virtual LANs, because its response to a client's configuration request (i.e., request to be configured) is the ATM address of an individual LAN Emulation server. Each LAN Emulation server is responsible for emulating a logical network and thus is aware of all the hosts that are members of that network. The LECS can be seen as the central control point for assigning hosts to their virtual LAN. By maintaining a database that contains end station address mappings corresponding to different LAN Emulation ARP servers, the LECS is capable of subdividing one physical ATM network into multiple logical networks.

11.3.2 LAN Emulation Server

The LAN Emulation Server (LES) is the focal point of the ARP process and it acts as the central repository of address mappings from ATM identifiers to MAC identifiers. Once a client has been assigned to an LES by the configuration server, it first creates a switched virtual circuit to the LES by generating a call setup with the destination being the LES's ATM address.

Once the call has been accepted by the LES, the client registers its ATM and MAC addresses with the LES. The client may also register other LAN destinations for which it is a proxy agent, as would be the case if the client were acting as a legacy bridge. In these situations, it may register several addresses on behalf of the devices attached to the legacy network(s).

Once the process of configuration and registration has been completed, the LES now contains a table listing the MAC addresses and the ATM addresses for all the hosts that are part of the emulated network. The clients then use the LES's table as a means to mimic the behavior of ARP. This is done by using the broadcast server, to send a conventional ARP packet to all the hosts on the network.

The ARP packet will contain the layer 3 address (i.e., IP, IPX) the client is trying to locate. When the reply is received from the target host possessing the layer 3 addresses, the source will have the MAC address of the destination. The source can then use the LES's address table to determine the correct mapping of the MAC address to an ATM address. The ATM address returned by the LES is then used by the source client to establish a direct virtual circuit between the source and the destination computers.

The key point to remember from this process is that the act of implementing the address resolution protocol over ATM has been hidden from the network layer applications. The combination of the different LANE servers has removed the complexities of ATM from the process of interhost communication.

11.3.3 LAN Emulation Broadcast and Unknown Server

As briefly described above, the Broadcast and Unknown Server (BUS) is the final component used to make a connection-oriented network appear connectionless. The BUS services are typically provided on a router, or other computer along with the distinct LANE servers, and are co-resident with the LES. The BUS's primary function is to emulate the mass distribution behavior of subnetwork broadcast—that is, any data sent to the broadcast address will be retransmitted to all the hosts on the emulated network.

Address resolution packets or video conferencing are examples of data types that are typically broadcast to all the hosts on a network.

To accomplish emulated broadcast, each host on the emulated LAN establishes a virtual circuit to the BUS serving its emulated LAN at registration time. With this virtual circuit the client, in much the same way as with the LES, joins the BUS's broadcast group by supplying the BUS with its ATM address. The BUS can then use this virtual circuit to send broadcast data that it has received from other clients that have similarly registered themselves. When data to be broadcast, such as address resolution messages, are sent to the BUS, it then has the responsibility of forwarding those data to all hosts in the group, thus completing the final component of emulating a LAN such as Ethernet or Token Ring.

11.3.4 LAN Emulation Clients

The LANE servers previously described act as the clearinghouse for messages and interoperate with LAN Emulation's fourth component: the actual devices attempting to communicate on the ATM network, known as LAN Emulation Clients (LECs). LEC is a computer or bridge attached to an ATM network running the LANE software. The tasks of the client-side software can be divided into two roles:

- Present a familiar/traditional layer OSI data link layer interface to the OSI network layer process running on the client
- Interact with the LAN Emulation servers to correctly implement the protocol and realize the address resolution function

Therefore, the LANE software residing on host computers or bridges performs data forwarding, address resolution, and control functions to emulate Ethernet or Token Ring. On a bridge, it represents, or proxies for, all the legacy-attached devices and acts as their agent when communicating with the LECS, LES, and BUS.

11.4 LAN Emulation operation

A host or bridge connected to an ATM network that LANE services is said to be connected to a LAN Emulation UNI (LUNI). In order for this interface to provide ARP-like functions, a set of requirements and procedures has been defined that each LEC must execute sequentially. In a LAN Emulation network, it is the operation of, and compliance with, the LUNI that allows all member components to correctly interoperate. (See figure 11.4.)

The processes of LANE can be subdivided into four areas. Each of these processes can be further subdivided into a virtual circuit establishment phase and an information exchange phase.

1 Initialization/configuration involves the LEC identifying itself to the LECS by presenting its ATM/MAC addresses and desired LAN type. In response to the registration request, it is given the ATM address of the LES responsible for the logical network, along with the emulated LAN type (i.e., Ethernet or Token Ring).
2 Registration/joining with the LES/BUS occurs when the LES establishes a virtual circuit to the LES: on this virtual circuit the client registers its ATM address, MAC address(es), LAN information, and proxy indicator—that is, if the LEC is a bridge. Next, it will establish a virtual circuit to the BUS and join the BUS's subnetwork group.

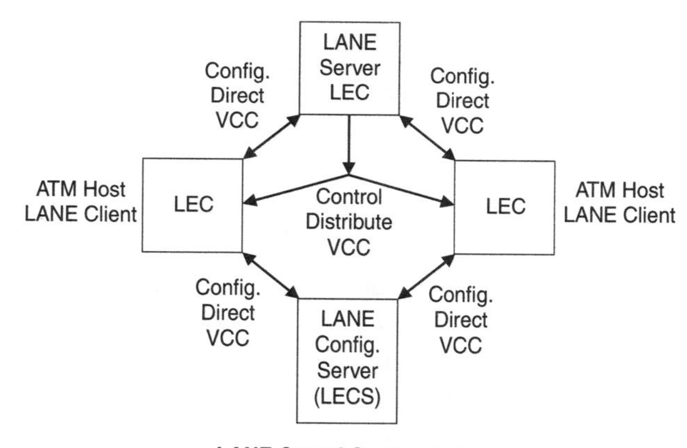

LANE Contol Connections

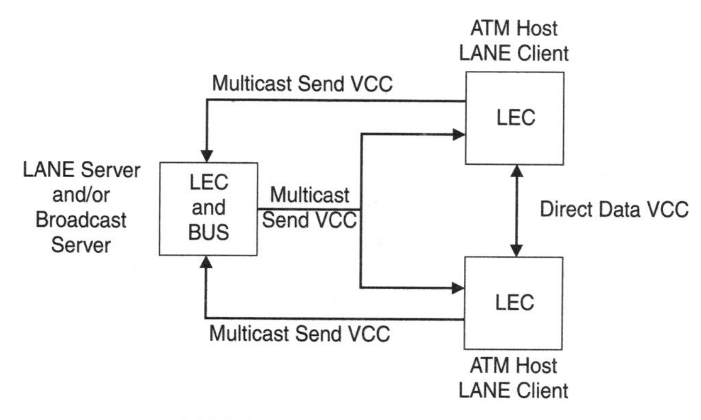

LANE Data and Multicast Connections

Figure 11.4 LAN Emulation UNI

3. Address Resolution using the LANE server: The LEC will transmit an ARP message, via the BUS, to all hosts on the emulated network. When the client receives a MAC address reply, it uses the LES to convert MAC to an ATM address.
4. Data Transfer with other LANE clients is accomplished by using the ATM address acquired from the LES to create a direct virtual circuit between the client and the destination.

The first step in using the LANE services is to establish a configuration-direct virtual circuit to the configuration server. This virtual circuit is called *direct*, because it refers to a point-to-point circuit between two hosts used only for direct communication between those hosts. In this phase of the protocol, the client establishes the virtual circuit to the LECS, identifies itself, and then receives a reply from the LECS assigning it to a LES. In some cases this step (getting a host up and running on a LANE network) can be omitted if the client is statically configured with the ATM address of the LES. In that case the client will boot up and then directly attempt to establish a virtual circuit to the LANE server that covers the client's domain.

Assuming that the client has successfully established a virtual circuit to the configuration server and acquired the ATM address of its server, it now must establish a virtual circuit to the LES and enter the "join state." In the join state, the LEC is assigned a unique LEC identifier (LECID), and told both the type of emulated LAN it is joining and the maximum frame size for intra-LAN communication. The LECID is used by the LES to determine where client messages are coming from and to perform error checking. Once the client has joined the emulation server, the LEC then registers at least one MAC address with the server. The server will use the LECID in these messages to determine the ATM address associated with the MAC addresses. Subsequent to registering the initial address, the LEC can optionally register additional MAC addresses for the devices it is bridging onto the emulated LAN.

Initialization is complete when the LEC performs its first ARP to locate and connect to the BUS. The LEC issues a LAN Emulation ARP message (LE_ARP) to the LES to request resolution of the BUS's address. In essence, the LEC is looking for an address mapping pair of MAC/ATM cached by the LEC in which the MAC portion is the emulated LAN broadcast address. The LES will reply to the LE_ARP with an LE_ARP_REPLY message containing the ATM address of the BUS. The LEC then establishes a virtual circuit to the BUS and registers itself as a member of the logical subnetwork with the BUS. At this point, the BUS establishes what is known as a unidirectional forwarding virtual circuit to the LEC. The forwarding virtual circuit is used to redirect all broadcast traffic received by the BUS back out at the clients on the emulated LAN. Once the forwarding virtual circuit establishment is complete, the initialization process is finished, and the LEC can begin to communicate, via ARP, with hosts other than the LANE servers.

When one client has data to transmit to a client whose ATM address is unknown, the originating client first transmits an ARP message to the BUS and then the message is forwarded to all hosts on the subnetwork. By implementing the standard ARP protocol, the desired destination will respond to the ARP request message with its MAC address. At this point the source can now generate an LE_ARP request and transmit it to its LES. The LE_ARP request contains the destination's MAC address and the product of this request will be a reply containing the MAC/ATM pair from the LES cache. After the LE_ARP_REPLY from the LES is passed to the client, the client can then use it to establish a direct virtual circuit to carry user data.

11.4.1 Multicast and unicast address forwarding

As described previously, after the process of address resolution is complete, clients can establish switched virtual circuits to destinations and use the new virtual circuits to exchange data between the destinations. Typically, when one computer wants to transmit a message to another, it will internally buffer the packets until it has completed the LE_ARP process. However, there is an alternative to holding the data that the clients can implement for short periods of time. If the originating host does not yet have a virtual circuit established to the destination, it can begin communicating prior to establishing the circuit if it transmits data to the destination via the BUS. The BUS then forwards the data to all the computers on the emulated LAN, where the true destination will receive the packets. This process is somewhat wasteful of network bandwidth, but it can be used to quickly disseminate data without creating virtual circuits.

Forwarding frames to all registered clients is acceptable and desirable behavior in an emulated LAN, due to the connectionless nature of the bridged networks being emulated. When the destination is unknown, as in an 802.1D transparent bridge [8], the frames reach their destination by being forwarded to every possible destination. The flooding process is critical to the correct operation of bridged networks, but it is also one of the major weaknesses of bridge networks and subsequently is a weakness of LANE when scaled to large internets.

If a client elects to use the BUS to flood data throughout the emulated network, it is recommended that it stop the forwarding once it has received an LE_ARP response to its original query. When the LE_ARP reply is received, the originator should establish a direct virtual circuit to the destination. If the client wishes to send multicast frames, then it must use the BUS's forwarding capability. This is a logical approach, because the BUS has established virtual circuits to all hosts on the emulated LAN. An LEC could exercise this option for broadcast in favor of establishing a mesh of virtual circuits to all possible destinations. In the case of unicast frames being transmitted between two hosts, the originating host should stop using the BUS as soon as it has established a direct-switched virtual circuit to the destination.

The final step used when emulating a LAN involves a protocol for purging data potentially buffered in the BUS during the ARP stage. The originating host can issue a request, which is sent to the BUS to expedite the transmission of all frames received from the originator. This process is called "flushing". Flushing is designed to prevent data from being delivered out of order. Without flushing, the originator may receive a response to the LE_ARP, establish a direct virtual circuit, and begin transmitting additional frames while data are pending transmission in the BUS. With flushing, the originator can establish a direct virtual circuit, issue a flush, momentarily pause, and then begin communicating with the destination.

11.4.2 Scalability and reliability

A key issue in developing a network based on LANE will be its performance capabilities related to scalability and reliability. In terms of scale, LANE, by design, is intended to emulate the operation of a legacy LAN segment. As bridged LANs grow, the number of broadcasts generated by ARP can become overwhelming. An additional problem is posed when the network is used to carry broadcast/multicast traffic—for example, if a LAN network is used for video distribution, it is conceivable that the BUS could quickly become overloaded. Both of these conditions have the potential to overload LAN Emulation servers.

To date, limited experience has been gained with large-scale production LANE networks and that experience has shown the protocol to be sound but the ultimate size of the networks is unknown. Scaling problems may be encountered, since packets must leave a subnetwork via a router that is a member of two or more emulated LANs. Most of the early deployment of LANE has focused on test beds and making single points of attachment to legacy LANs—for example, using a LAN Emulation edge device to connect the ATM backbone to a campus Ethernet or Token Ring (see figure 11.5).

LANE does provide the benefits of far superior line speeds coupled with a switched environment. Typically, hosts connected to a LANE network support interface speeds of 155 Mbps. Since LANE is not tied to any physical medium, it is potentially scalable from 64 Kbps to OC-198. More realistically, it is possible to attach LANE servers to networks via 622 Mbps connections that further reduce congestion and improve performance. These high speeds may facilitate a migration back to centralized servers capable of providing information to thousands of clients. When properly implemented, the network can statistically multiplex best-effort traffic, thus providing very high speed inter-host communication channels with little congestion.

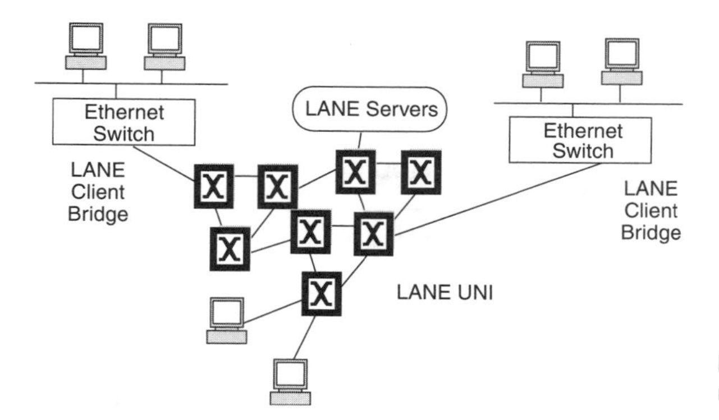

Figure 11.5 Small scale LANE network

11.5 Classical IP over ATM overview

In the previous section, the problem of running LAN traffic over ATM was addressed with the LAN Emulation suite of protocols. In this section, an alternate technique, known as Classical IP over ATM, will be discussed. Classical IP over ATM (CIP) slightly predates LANE and is the method of running LAN traffic over ATM developed by the Internet Engineering Task Force (IETF). The IETF's specification is defined to provide native IP support over ATM and is documented in the following requests for comments:

- RFC 1483, "Multiprotocol Encapsulation over ATM Adaptation Layer 5"
- RFC 1577, "Classical IP and ARP over ATM"
- RFC 1755, "ATM Signaling Support for IP over ATM"
- RFC 2022, "Multicast Address Resolution (MARS) Protocol".

These protocols are designed to treat ATM as a virtual *wire* with the special property of being connection-oriented, and, therefore, as with LANE, requiring a unique method for address resolution and broadcast support.

In the Classical IP over ATM model, the ATM fabric interconnecting a group of hosts is considered a network, called Non-Broadcast Multiple Access (NBMA) [9]. An NBMA network is made up of a switched service such as ATM or frame relay, with a large number of end stations that cannot directly broadcast messages to each other. While on the NBMA network there may be one OSI layer 2 network; it is subdivided into several Logical IP Subnetworks (LISs) [6,7] that can be traversed only via routers.

One of the main philosophies behind Classical IP over ATM is that network administrators will build networks using the same techniques in use today—that is,

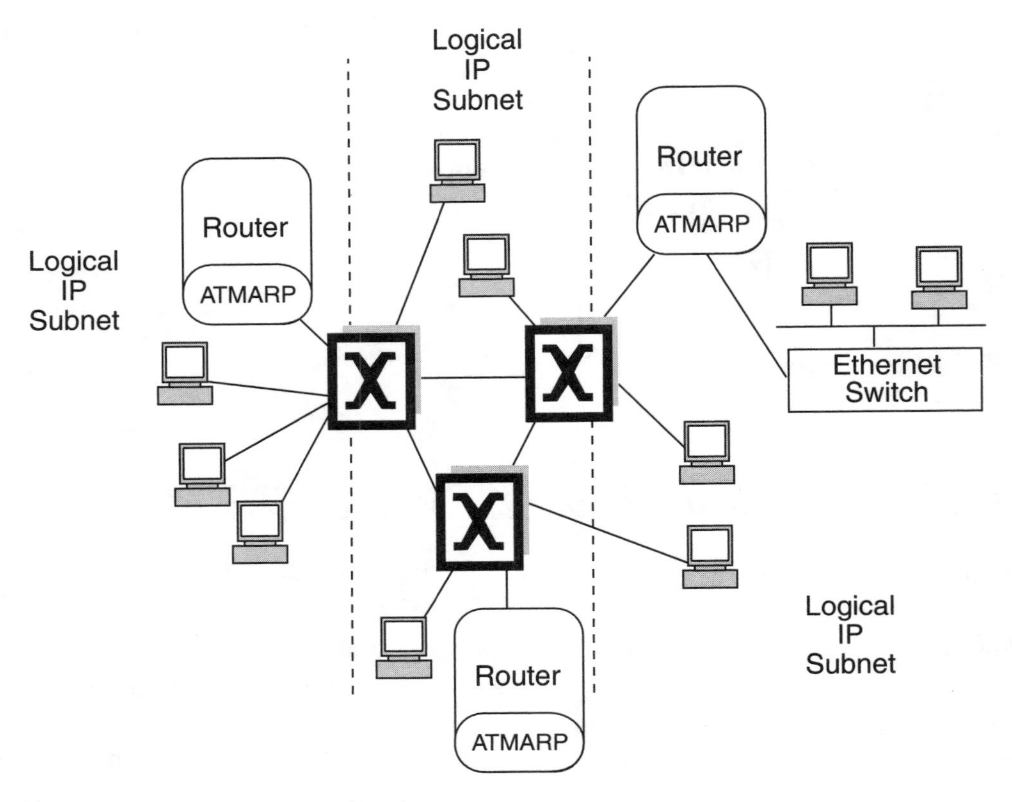

Figure 11.6 Classical IP over ATM LIS

dividing hosts into physical groups, known as subnetworks, according to administrative workgroup domains. Then the subnetworks are interconnected to other subnetworks via IP routers. An LIS in Classical IP over ATM is made up of a collection of ATM-attached hosts and ATM-attached IP routers that are part of a common IP subnetwork. Policy administration, such as security, access controls, routing, and filtering, will still remain a function of routers, because the ATM network is just "smart wire."

In Classical IP over ATM, as with the LANE protocol, the functionality of ARP is provided with the help of special-purpose server processes typically collocated on the router. This is done via software upgrades with the legacy IP routers. Each Classical IP over ATM LIS has an ARP server that maintains IP address-to-ATM address mappings. All members of the LIS register with the ARP server, and subsequently all ARP requests from members of the LIS are handled by the ARP server. This mechanism is a little more straightforward than LANE version 1, since, for ARP, there is only one server and this server maintains direct IP-to-ATM address mappings. (See figure 11.6.)

In the CIP model, IP ARP requests are forwarded from hosts directly to the LIS ARP server using MAC/ATM address mappings acquired at CIP registration. The ARP server, running on an ATM-attached router, replies with an ATM address. When the ARP request originator receives the reply with the ATM address, it can then issue a call setup message and directly establish communication with the desired destination.

As opposed to LANE, the CIP model's simplicity reduces the amount of broadcast traffic and interactions with various servers. By reducing communication with the LECS, LES, and BUS, the time required for address resolution can be reduced. In addition, once the address has been resolved, there is a potential that subsequent data transfer rates may be reduced. However, the reduction in complexity does come with a reduction in functionality. CIP has the drawback of supporting only the IP protocol, because the ARP server is only knowledgeable about IP. The base CIP protocol also has no understanding of QoS.

As with LANE, communication between LISs must be made via ATM-attached routers that are members of more than one LIS. One physical ATM network can logically be considered as several LISs, but the interconnection, from the host perspective, is accomplished via another router. Using an ATM-attached router as the path between subnetworks, prevents ATM-attached end stations in different subnetworks from creating direct virtual circuits between one another. This restriction has the potential to degrade throughput and increase latency.

There are also questions about the reliability of the IP ARP server, since the current version of the specification has no provisions for redundancy. If the ARP server were to suffer a catastrophic failure, all hosts on the LIS would be unable to use the ARP. Finally, Classical IP over ATM suffers from the drawback of each host needing to be manually configured with the ATM address of the ARP server, as opposed to the dynamic discovery allowed in LANE.

11.5.1 Classical IP Multicast

Classical IP provides multicast support via the Multicast Address Resolution Server (MARS) [10]. The MARS model is similar to a client/server design because it operates by requiring a Multicast Server (MCS) to keep membership lists of multicast clients that have joined a multicast group. (See figure 11.7.) A network administrator assigns a client to a multicast server at configuration time. In the MARS model, a MARS system along with its associated clients, is called a *cluster.* The MARS approach uses an address resolution server to map an IP multicast address from the cluster onto a set of ATM end-point addresses of the multicast group members.

The MARS model is based on a hierarchy of devices, as shown in figure 11.8. The three primary components of a MARS-based IP over ATM network are as follows:

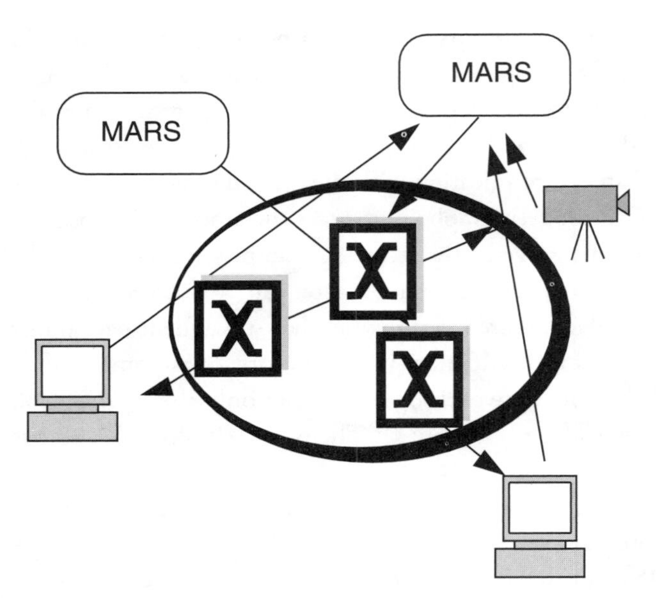

Figure 11.7 Multicast with MARS and MCS

- Top level server(s) known as the MARS
- Zero or more multicast servers that provide second-level multicast distribution
- Clients that utilize IP multicast by building point-to-multipoint paths based on information learned by MARS

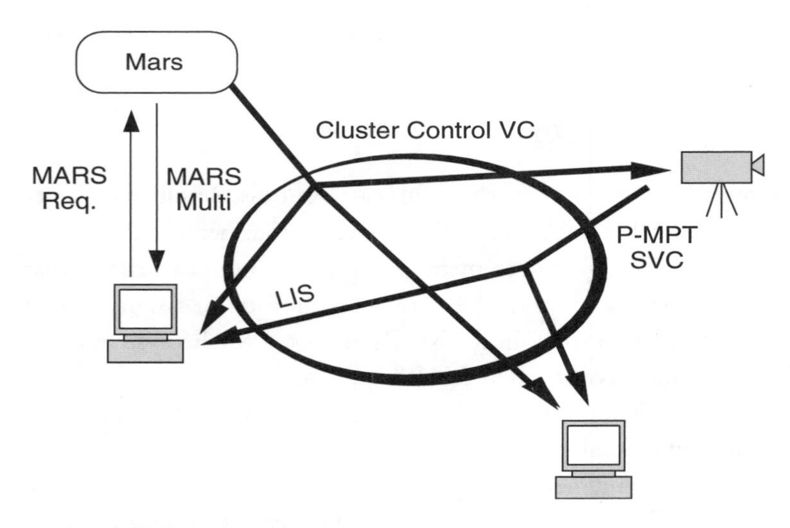

Figure 11.8 MARS message flows

Every MARS* has at least one client and server contained within a Classical IP over ATM Logical IP Subnetwork. To operate the system, clients use the MARS as a means of determining what other hosts are members of a multicast group. In a MARS network there are two modes of operation: full mesh or multicast server.

In the full-mesh mode, client queries are sent to the server to identify which hosts have registered as members of a Class D tree. A Class D address is part of the global IP multicast range. Next, the client establishes a point-to-multipoint virtual circuit to those leaves. In the multicast server mode, the MCS acts as the focal point of all multicast packets originated anywhere in the multicast tree. In this case, in order to simulate IP multicast over an ATM network, the MCS simply retransmits, over the ATM multicast connections, all packets sent to the IP multicast group by the clients. Because the set of hosts in a multicast group is constantly changing, the MARS is also responsible for dynamically updating the set of clients with new membership information as changes occur as well as adding and removing clients.

When running multicast over an ATM network, selecting between the two modes of operation just described is left to the discretion of the network designer. Adding multiple layers of hierarchy to the distribution tree can call for an additional design determination with MARS. For example, multicast clusters may contain the second level of the hierarchy by elevating a client to the roll of a multicast server.

One of the tradeoffs conceded in order to gain the simplicity that MARS offers is the required "out of band" control messages used to maintain multicast group membership. In order for clients and multicast servers to send and receive control and membership information, the MARS protocol specifies the setup of a partial mesh of virtual circuits. The MARS maintains its own point-to-multipoint circuits, the ClusterControlVC, for the members within the cluster.

The ClusterControlVC carries leaf node update information to the clients as members leave and join the multicast session. Each client in a multicast cluster maintains a point-to-point virtual circuit to the MARS that is used to initialize itself and for path group change messages. Finally, the MARS manages the MCSs through point-to-point virtual circuits between each MCS and the MARS and point-to-multipoint circuits, called ServerControlVC, from the MARS to the MCSs. These circuits are used to pass information from the MARS to keep the cluster membership updated.

The MARS protocol has defined a set of control messages that are exchanged between the MARS and the clients in order to maintain the group memberships.

- MARS_JOIN/LEAVE: These messages are used by clients to register with the MARS as cluster members. The messages are also used to join or leave particular IP

---

* Typically, the MARS is collocated with the CIP ARP server.

multicast groups. These messages carry the ATM address of the client joining/leaving and the Class D multicast IP address. When the message is sent to the MARS without an IP address, it is assumed to be a cluster registration. When a MARS_JOIN/LEAVE message is received by the MARS, it is propagated on the ClusterControlVC to the members of the cluster. Clients that receive the message then must establish (or delete) a point-to-multipoint circuit to the new node.

- MARS_REQUEST: These messages are issued by clients when they first join a MARS. The message is sent to the MARS to determine the set of hosts that are part of a multicast group. The MARS responds by either notifying the client that no hosts are in the session or producing a list of the session's hosts. The former response takes the form of a MARS_NAK message, and the latter response is known as a MARS_MULTI.
- MARS_REDIRECT_MAP: These messages act as a "heartbeat" mechanism ; they provide a steady sanity check of the MARS system. The messages are periodically transmitted from the MARS to the clients over the ClusterControlVC. The MARS architecture supports redundancy of MARSs. Inside the MARS_REDIRECT_MAP is a list of one or more ATM addresses representing the current and backup MARSs. Typically, the MARS_REDIRECT_MAP is examined to determine which MARS should be used in the event the primary fails. If a client fails to receive a user-specified number of MARS_REDIRECT_MAPs messages, it can then assume the MARS has failed and it should use the secondary. In addition, the messages can be used to signal a graceful cutover to a new server. If a client receives a MARS_REDIRECT_MAP in which the primary MARS has changed, then the client is expected to modify its ClusterControlVC to the new MARS.
- MARS_MIGRATE: Messages are used to force the deletion and subsequent recreation of a client's point-to-multipoint circuits for a multicast group and replace them with new circuits, however, the new circuits should be established to the hosts contained in the MARS_MIGRATE payload list. This message is used to force the migration from a mesh-based virtual circuit distribution to a multicast server-based distribution.

In addition to the message exchanges taking place between the MARS and clients, as just discussed, the MARS has a special set of messages that it exchanges with the second-tier MCSs (should one exist).

- MARS_MSERV/UNSERV: These message types are exchanged between an MCS and the MARS to signify the MCS's ability to begin/end being an MCS for a particular Class D multicast session.
- MARS_SJOIN/SLEAVE: These messages are propagated on the ServerControlVC by the MARS when it is necessary to inform the MCSs that their point-to-multipoint shared forwarding VCs may need updating. An MCS receiving one of these messages must establish/delete a virtual circuit the client has specified in the message payload.
- MARS_REQUEST: These messages are issued by MCSs when they are first registering with the MARS to determine the set of hosts that is part of a multicast group. The MARS responds by notifying the MCS that either there are no hosts in the session, MARS_NAK, or it responds with a list of the session's hosts in a MARS_MULTI.
- MARS_REDIRECT_MAP: Similar to the client, the MCS also receives a MARS_REDIRECT_MAP which determines the ATM address of a second MARS to use in the event of failure.

The following list indicates the steps followed by clients wishing to exchange IP multicast traffic on a MARS system.

1. Register with the MARS
   - Transmit a MARS_JOIN with only the client's ATM address.
2. Multicast Group Join
   - Transmit a MARS_JOIN containing one or more Class D IP multicast addresses and the client's ATM address.
   - The MARS then propagates the MARS_JOIN to all clients that have already joined that multicast group.
   - Each sender that receives the MARS_JOIN adds the new client to its ATM multicast connection.
3. Data Transmission
   - Transmit a MARS_REQUEST to the MARS containing the Class D IP multicast address and the ATM address of the client.
   - The MARS then returns a list of ATM addresses associated with the IP multicast address in a MARS_MULTI, or, if there are no receivers in the groups, responds with a MARS_NAK.

11.5.2 Classical IP operation

In the Classical IP model, data transfer is done by creating a virtual circuit between hosts and then using LLC/SNAP encapsulation of data that have been segmented by AAL5. Mapping IP packets onto ATM cells using LLC/SNAP is specified in RFC#1483, "Multiprotocol Encapsulation over ATM". RFC#1483 specifies how data are formatted prior to segmentation (see figure 11.9). The RFC documents several different methods; however, the vast majority of host/router implementations use the LLC/SNAP encapsulation. LLC/SNMP specifies that each datagram is prefaced with a bit pattern, which the receiver can use to determine the protocol type of the source.

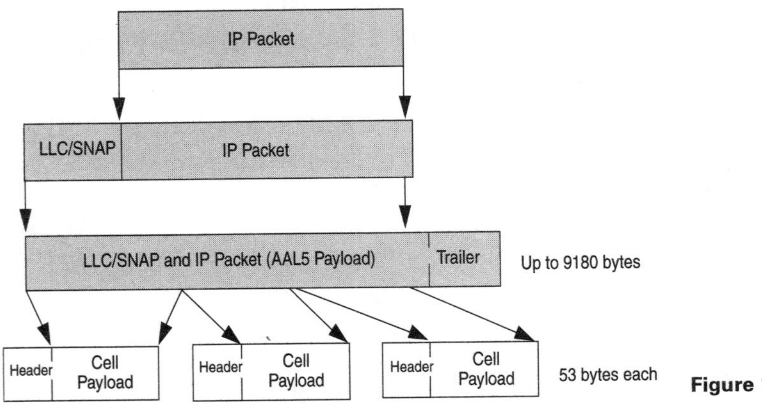

Figure 11.9 RFC 1483

The advantages provided by the encapsulation method specified in RFC#1483 are that it treats ATM as a data link layer that supports a large Maximum Transfer Unit (MTU) and that it can operate in either a bridge or multiplexed mode. Because the network is not emulating an Ethernet or Token Ring, such as LANE, the MTU has been specified to be as large as 9,180 bytes. The large MTU can improve performance of hosts attached directly to the ATM network.

RFC 1577 specifies two major modifications to traditional connectionless ARP. The first modification is the creation of the ATMARP message used to request addresses. The second modification is the InATMARP message, which inverses address registration. When a client wishes to initialize itself on an LIS, it establishes a switched virtual circuit to the CIP ARP server. Once the circuit has been established, the server contains the ATM address extracted from the call setup message calling party field of the client.

The server can now transmit an InATMARP request in an attempt to determine the IP address of the client that has just created the virtual circuit. The client responds to the InATMARP request with its IP address, and the server then uses this information to build an ATMARP table cache. The ARP table in the server will contain a listing for IP-to-ATM pairs for all hosts that have registered and periodically refreshed their entries to prevent them from timing out. The ATMARP server cache answers subsequent ATMARP requests for the client's IP address. Clients wishing to resolve addresses generate ATMARP messages, which are sent to their server, and locally cache the reply. Client cache table entries expire and must be renewed every 15 minutes. Server entries for attached hosts timeout after 20 minutes.

11.6 Integrating ATM signaling

In this section, the description of CIP will be expanded upon by a discussion of how the protocol is mapped to ATM signaling messages. In addition, the chapter will conclude with a discussion of the mapping from the perspective of protocols in a nonintegrated service model, which is a topic developed by both the ATM Forum and the IETF [11]. The primary focus of their work is in mapping IP to ATM best-effort service; however, some work focuses on the interaction of TCP with ABR [3]. This information is important, because it helps describe how legacy protocols can be used on ATM networks in their most basic form. Once this basic implementation is in place, we will describe in the following chapter the next steps in employing MPOA to explore QoS and shortcut routing.

When mapping a TCP connection to a switched virtual circuit, the client creates an ATM UNI call setup message, which conforms to standards developed by the ATM Forum. It then transmits the message to its local ATM switch. Most call setups follow a client/server-style message exchange, as shown in figure 11.10. In the figure, the host on the far right is attempting to establish a circuit to the other host. The call setup message is generated and then passed through the interconnecting switches, where is it routed by PNNI and ultimately reaches the destination switch.

The source initiates creation of a circuit by sending a SETUP message. The local ATM switch responds to this message with CALL PROCEEDING, which contains the call identifier and the virtual path identifier/virtual circuit identifier (VPI/VCI) to use for the duration of the call. The call identifier is an integer that the client uses to refer to the virtual circuit in subsequent signaling messages. When the destination accepts the call, it transmits a CONNECT to its switch. When the CONNECT propagates back to the originator, the virtual circuit creation is complete and ready for use.

The call SETUP message is made up of a string of Information Elements (IEs), as defined by the ATM Forum, that define the parameters of the switched virtual circuit. There is both a set of mandatory IEs, such as destination address, and others, such as

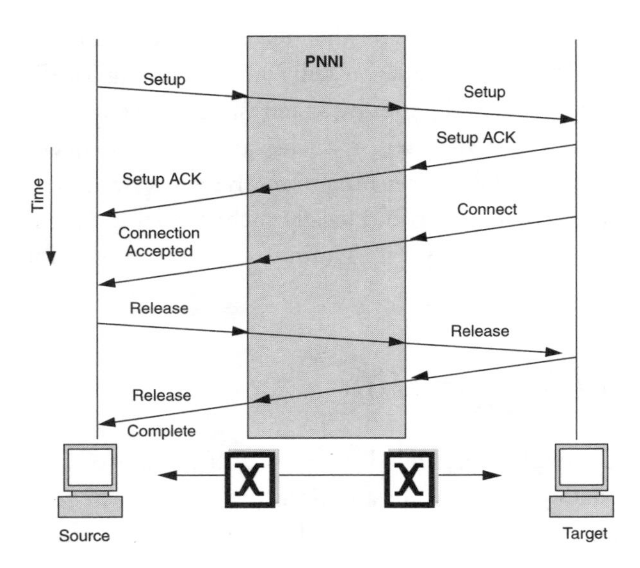

Figure 11.10 Establishing a switched virtual circuit

transit carrier, that are not always required. Any host, or the network ATM switches, can release calls when it generates a RELEASE message. At the end of a file transfer, for example, the source may issue the RELEASE, with the appropriate call identifier, to start the process of tearing down the circuit. The RELEASE message will cause the interconnecting devices to generate RELEASE COMPLETE, which signifies that the circuit has been deleted.

An area of complexity often encountered when building ATM networks is compatibility and support of signaling in the switches and hosts. There are three signaling standards—UNI 3.0, UNI 3.1, and UNI 4.0—that have been released by the ATM Forum and are in use today. The first and second protocols do not interoperate, due to protocol changes above the AAL, and the second and third protocols differ by the services and features supported. UNI 3.0/3.1 provides basic switched virtual circuit support with root-controlled multicast trees. The UNI 4.0 specification builds upon UNI 3.1 with the addition of several interesting features: ABR, traffic parameter negotiation, frame discard, switched virtual path, and improvements to multicast with Anycast and Leaf-Initiated Join (LIJ).

ATM signaling messages are complex and their construction can take many different forms. In order to simplify the process of signaling and ensure agreement among the vendors, the IETF has released a document, RFC#1755, setting the baseline of IEs to be used and also which parameter values should be used, when possible, for specifying default IEs. When using ATM signaling to establish a point-to-point connection, the mandatory IEs for UNI 3.1 messages, as specified by the IEFT, are as follows:

- AAL parameter: This specifies which AAL will be used. AAL5 must be indicated in this field.
- ATM traffic descriptor: This describes the traffic using ATM's traffic parameters: PCR, SCR, and MBS. It is also possible to supply the service category with this parameter. RFC 1755, a protocol that does not use integrated services, recommends best-effort service. If, when using best-effort, the message fails with the error code "User Cell Rate Unavailable," the call may be retryed with a smaller value of the PCR. The RFC also recommends that the PCR be set to 1/10 of the line rate.
- Broadband bearer capability: This field can be set to either connection-oriented service type X (unspecified) or type C# (connection-oriented). In most cases, best-effort should use type X.
- Broadband lower layer information: This field is used to specify the encapsulation of the data and, in keeping with RFC 1483, this value should be set to LLC as the default.
- QoS parameter: "Unspecified QoS" is the only QoS class that must be supported by all networks and allowed when using the best-effort service, as specified in RFC 1755.
- Called party number: This is ATM addressing information of the destination.
- Calling party number: This is ATM addressing information of the source*.

Optional UNI 3.1 IEs include the following.

- Calling party sub-address: This ATM addressing information usually associated with a LAN address with the primary ATM address used for WAN identification.
- Called Party Sub-address: ATM addressing information is usually associated with a LAN address with the primary ATM address used for WAN identification.
- Transit network selector: When appropriate, this field selects an interexchange carrier.

When using the UNI 4.0 signaling specification, there are new IEs required to control the extra features. When reading the UNI 4.0 signaling additions, it seems as if the ATM Forum focused its attention on making ATM more usable for TCP/IP data applications. There are several new and useful features in the UNI 4.0 signaling document

---

* The source can learn about its own address using the Interim Local Management Interface (ILMI) protocol, as specified in section 5.8 of UNI 3.1

for mapping TCP/IP to ATM switched virtual circuits. The 4.0 specification supports these additional features.

- Traffic parameter negotiation: This allows the SETUP to contain multiple IEs for the same object with the intent of reducing CAC failures. If the first IE is unacceptable for call completion, the switch has the option of retrying the CAC with the second IE.
- Available bit rate: This utilizes ATM's closed loop flow control by requesting in the signaling message that the service be associated with the new circuit. ABR also functions as a means of traffic parameter negotiation, because it allows the user to establish a baseline and then request modification after the circuit is in service. Applications can learn from the network how much data can be transmitted per second.
- Multicast extensions:
  - Leaf-Initiated—Join (LIJ) enables clients to add themselves to a multicast tree without the root's knowledge.
  - Anycast—allows one ATM address to correspond to multiple receivers.
- Virtual path switching: This permits the signaling of an entire virtual path instead of the usual VPI/VCI granularity.
- Frame discard service: This allows signaling to request that this service be associated with a circuit, so that during congestion the Partial Packet Discard (PPD) algorithms are employed [12].
- Signaling of individual QoS parameters: The UNI 4.0 specification diverges from the UNI 3.1 philosophy by selecting a QoS service category.

With these new features comes a set of new IEs that evokes and controls their behavior. The new IEs used when establishing virtual circuits on a UNI 4.0-capable switch are as follows.

- Minimum acceptable ATM traffic descriptor: This field is used with the ABR service category. It sets the baseline for the ABR service and specifies the lowest bit per second that can be transmitted. ABR setup parameters include the various objects used to initialize ABR:
  - Initial cell rate
  - RM roundtrip time
  - Data rate increment factor

    - Data rate decrement factor
    - Transient buffer exposure

- Alternative ATM traffic descriptor: This field is used with the new feature of multiple IEs for the same parameter. If the ABR service is being selected, the alternative traffic descriptors are prohibited.
- The ATM traffic category (CBR, VBR-nt, etc.) is selected by a new field, `transfer_capability`, in the broadband bearer capability IE. This IE also specifies whether the signaling message is point-to-point or multicast.
- Connection scope selection.
- Extended QoS parameters: These are used to specify the values of cell delay variation and cell loss ratios. The QoS parameters are not directly specified by the IEFT.
- End-to-End transit delay.

When mapping CIP to ATM signaling, the application must choose whether it will run the TCP session over an ABR circuit or Partial Packet Discard (PPD) circuit. The mode must be decided upon prior to creating the virtual circuit, because the set of IEs carried in the signaling message is different. Setting the frame discard bit in the traffic management option subfield of the traffic descriptor IE does the signaling for frame discard. To date, most IP VCCs use PPD.

11.6.1 IP multicast over UNI 4.0 signaling

The new features supported in UNI 4.0 signaling can also be used to modify the operation of IP multicast with MARS. However, the two MARS approaches differ in the responsibilities that sources and receivers have. In the UNI 3.1 model, the sources have total control over receivers and must build the multicast tree. In the UNI 4.0 model, the receivers can assume the role of tree construction.

As with the UNI 3.1 MARS model, each client wishing to use multicast IP builds a point-to-point virtual circuit, used to exchange control information, to the MARS. From there the model changes from RFC 2022 in that each IP multicast sender registers its ATM address along with the call identifier of a virtual circuit it has created. When the sender initially creates the multicast circuit, it specifies in the bearer capability IE that the circuit supports UNI 4.0 Leaf-Initiated Joins (LIJs). As additional clients wishing to be receivers register with the MARS, they are provided with the source's ATM address and the call identifier,which can be used to issue an LIJ.

When the MARS adds a client that is a sender for an IP multicast group, it broadcasts this information over the cluster control virtual circuit to all of the potential receivers in its domain. Receivers interested in the session can then issue LIJ call SETUP messages to their local switch, and they will be added to the multicast virtual circuit without the roots intervention.

The model described above has the potential of providing good scalability in environments where there is a small workgroup or in systems with few numbers of senders and large numbers of receivers. An example of the latter would be the distribution of broadcast television where it would technically be very difficult for the root to issue a unique message to add all receivers. There are, however, some concerns with UNI 4.0 LIJ multicast and IP multicast in general, and research has shown that in order to obtain high degrees of scalability, packet-level multicast servers are required [13].

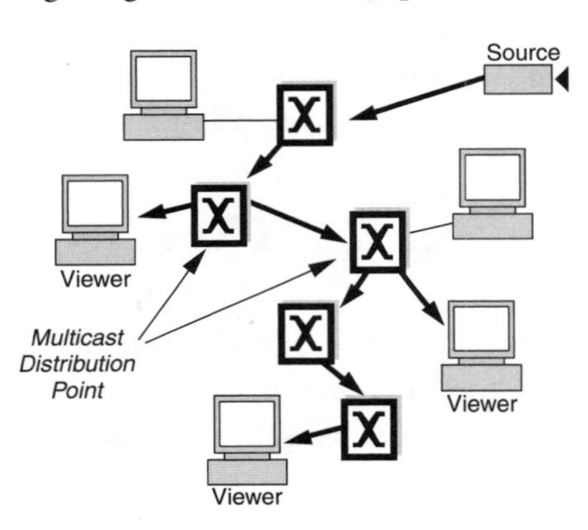

Figure 11.11 Multicast tree

The LIJ model does pose new security problems not encountered in UNI 3.1. Because the root is not aware of receivers being added, there is a potential for a security breach without the sender ever knowing. An additional problem has to do with scalability in a many-to-many multicast exchange over ATM. When IP multicast is transmitted over traditional IP multicast backbones, the interconnecting routers act as points of aggregation and simplify the multicast tree, as shown in figure 11.11.

The benefit of this aggregation property is that only a single packet needs to traverse most points of the network before it reaches a cluster of receivers. When large multicast sessions are run over a pure ATM network, the aggregation property is lost and must be replaced with several virtual circuits, as shown in figure 11.12.

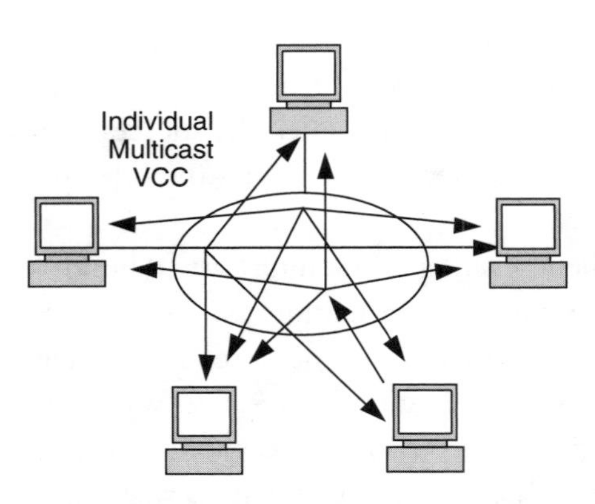

Figure 11.12 Multicast tree without routers

Therefore, in order to build networks that scale to very large sizes, it may be necessary to use some routers, or aggregation devices, that form points of concentration at the cluster boundaries. These devices would be used to reduce the virtual circuit count and the number of identical copies of packets being transmitted. In addition, the aggregation router provides the benefit of localized group membership changes (see figure 11.13.)

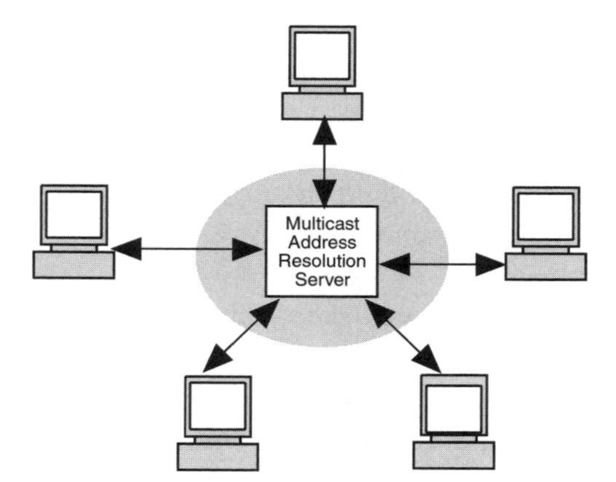

Figure 11.13 MARS Multicast tree with routers

11.7 *Summary of MPOA LAN communication*

This chapter has described methods and protocols used to immediately build ATM networks that form the basis of MPOA. The approaches can be subdivided into two groups.

1. The ATM Forum's LAN Emulation
2. The IETF's Classical IP/MARS over ATM

At the heart of these protocols is the desire to provide the functionality of ARP over an ATM network. The LANE protocol accomplishes this task by emulating the OSI layer 2 technologies, Ethernet and Token Ring. Classical IP over ATM provides a protocol that allows hosts to ARP via a server mapping IP to ATM. Finally, MARS allows ATM hosts to participate in IP multicast sessions.

Both approaches function well and are widely used. However, both protocols have their limitations. LANE suffers from potential scalability problems associated with all bridged networks. Classical IP over ATM suffers from a lack of support for several very popular network layer protocols, such as IPX and AppleTalk. Neither protocol makes use of routing between subnetworks using ATM as a shortcut, and both need conventional routers to interconnect different subnetworks. Finally, and possibly most importantly, neither clearly takes advantage of ATM's extensive QoS features.

Classical IP over ATM maps IP directly onto ATM, which eliminates some limitations of the LAN Emulation model because of reduced protocol encapsulation and operations overhead. With Classical IP over ATM, ARP requests are forwarded directly to the ARP server, which replies with an ATM address. This gives the originating station everything it needs to establish a virtual circuit to the target destination in a single step, instead of LANE's multimessage process. By cutting down on the number of steps needed to establish connections between the source and destination, the Classical IP over ATM approach minimizes broadcast traffic and improves latency. Another advantage of mapping IP directly over ATM, as opposed to emulating a legacy LAN, is the ability to use a larger MTU than would be possible on an emulated Ethernet or Token Ring.

Expanding upon the work presented in this chapter was the charter of the MPOA working group and in the following chapter we will discuss what MPOA brings to the table to address QoS features and the ability to determine the shortest path through complex ATM/IP networks. Chapter 12 will focus on integration of LAN APR mechanisms and will begin our discussion of the Next Hop-Resolution Protocol. This protocol paves the way for shortcut path selection in MPOA and lays the groundwork for documenting a complete system tying together the functionality described in the preceding chapters.

11.8 References

1. ATM Forum, *ATM User Network Interface Specification.* Englewood Cliffs, NJ: Prentice Hall, 1993.
2. Laubach, *Classical IP and ARP over ATM,* RFC 1577, located at *www.ietf.org* (January 1994).
3. W. Richard Stevens, *TCP/IP Illustrated.* Reading, MA: Addison-Wesley, 1994.
4. *ATM Forum, ATM User Network Interface Specification*, v4.0, located at *www.atmforum.com.*
5. *www.ipsilon.com.*
6. M. Rose, *The Open Book*, Englewood Cliffs, NJ: Prentice Hall, 1992.
7. D.L. Spohn, *Data Network Design*, New York: McGraw-Hill, 1993.
8. ISO/IEC, 10038 *ANSI/IEEE Std. 802.1D Information Processing Systems—Local Area Networks—MAC Sublayer Interconnection (MAC Bridges).*
9. D. Katz, D. Piscitello, and B. Cole, *NBMA Next-Hop Resolution Protocol (NHRP),* located at *ftp://ietf.cnri.reston.va.us/internet-drafts/draft-ietf-rolc-nhrp-07.txt* (December 29, 1995).
10. G. Armitage, *Support for Multicast over UNI 3.0/3.1 Based ATM Networks,* located at *ftp://ietf.cnri.reston.va.us/internet-drafts/draft-ietf-ipatm-ipmc-12.txt* (December 29, 1995).
11. M. Perez, F. Liaw, D. Grossman, A. Mankin, E. Hoffman and A. Malis, *ATM Signaling Support for IP over ATM,* RFC 1755, located at *www.etf.org* (February 1995).

12 *www.sigcom.org.*

13 Armitage, *VENUS—Very Extended Non-Unicast Service,* located at *http://ds.internic.net/internet-drafts/draft-armitage-ion-venus-03.txt.* (June 1997).

CHAPTER 12

Realizing Multiprotocol over ATM Networks

Throughout the previous chapters, the groundwork has been established regarding the tools and protocols used by data communications managers to build networks supporting legacy protocols over ATM and, ultimately, QoS. Building on this information, this chapter examines the work done by the ATM Forum with regard to the Multiprotocol over ATM (MPOA) protocol suite. MPOA is considered a protocol suite because its primary goal is not to develop new technology, but rather to draw upon the work of several different standards bodies synthesizing them into a "big picture" document, which describes how all these protocols interact.

MPOA provides a means for seamlessly internetworking a triad of protocols. On one side there are ATM's legacy protocols, such as Classical IP and LAN Emulation, that allow hosts to discover each other on a local logical subnetwork and form the basis for intra-LAN communication. On the second branch, there are protocols developed in the IEFT which allow hosts to establish direct communication paths across an ATM network traversing subnetwork boundaries without a router. The final leg of the triad is provided by a set of integrated service protocols, such as RSVP and its associated protocols RTP/RTCP/RTSP, which provide a means to specify QoS and then help realize and monitor network performance.

The MPOA specification, like many cutting-edge technologies, is a work in progress. However, the first phase of the specification is complete and MPOA's future directions are clear. The goal of the first phase of the MPOA specification is to document techniques for "shortcut" routing and integration of MPOA with LANE and legacy networks. The first major distinction between MPOA and its predecessors, LANE has the ability to support the Next-Hop Resolution Protocol (NHRP) [1] on the multiprotocol servers. Subsequent phases of MPOA will integrate QoS and multicast protocols and discuss how different protocols interact with an ATM network.

The material in the previous chapter about LAN Emulation and Classical IP over ATM illustrated the techniques that can make existing applications work on ATM networks. The major differences between those protocols and MPOA will become clearer as the phases of the work are described. In conjunction with a discussion of the MPOA specification, material on the Next-Hop Resolution Protocol, flow modeling, and legacy integration will be covered. This chapter also presents MPOA ideas that are being applied to network design and routing as well as how network managers can apply the MPOA concepts to build next-generation networks and applications.

12.1 MPOA's background

As described in the previous chapter, the techniques of LAN Emulation and Classical IP over ATM are only starting points for building ATM networks. In order to take full advantage of ATM's potential, new paradigms in network design and application development must be undertaken. MPOA is a product of this paradigm shift and may revolutionize the way networks are built and used. The paradigm changes are already beginning to be felt, as manufacturers release products that employ the MPOA architecture philosophies of separating switching from routing and allowing applications to designate their required QoS.

For network managers, MPOA can be viewed as a superset protocol, which brings together many underlying technologies and is responsible for solving the problems of establishing connections between pairs of hosts that cross administrative domains, as well as enabling applications to make use of a network's ability to provide guaranteed QoS. Some of the key advantages of MPOA include the following:

- Edge devices can establish direct connections between themselves without using routers
- Lower latency communication between devices after ARP can be accomplished due to route elimination
- There is a reduced/restricted amount of broadcast traffic
- There is flexibility in the selection of maximum transfer unit size to optimize performance.
- Fabrication of multiple virtual LANs can be done on one ATM network.

12.2 Virtual LANs

One of MPOA's main goals is to enhance the ability of network designers to build a single physical ATM network for enterprise networks, while, at the same time, providing the ability to subdivide the network along administrative boundaries (i.e., build virtual LANs). The concept of virtual LANs is a generic idea, where the end station's logical addresses (IP) are detached from their physical location (MAC). When a network is empowered with knowledge to subdivide itself into multiple segments via address servers, then the network is not just a transparent pipe but is knowledgeable about the connected devices. For this reason, the network is capable of dynamically supporting host movement and additional configuration changes, detecting duplication of addresses, and adding unauthorized addresses.

Dynamic movement with automatic reconfiguration can be a very useful feature to help network administrators cope with the problem of incorrectly addressed hosts disrupting correctly addressed hosts or detecting when a malicious user has attached a device to the network in an attempt to eavesdrop. By creating virtual LANs, the ability

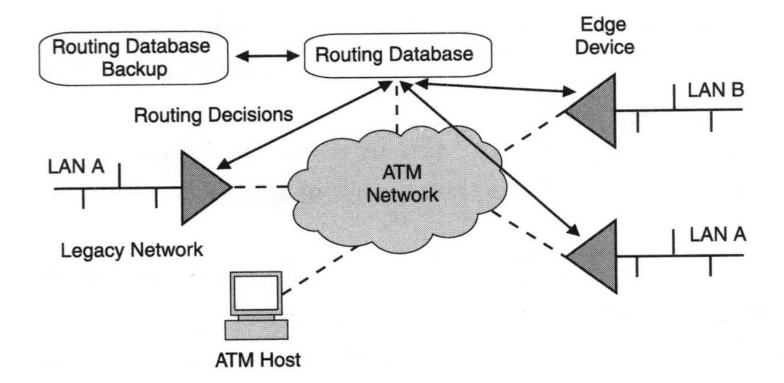

Figure 12.1 Subdivision of routing tasks

to eavesdrop can also be reduced. Consequently, security can be much higher than on conventionally shared media. More importantly, the ability to place configuration information into a central server allows the campus network to become more dynamic and easier to manage. A large percentage of the costs associated with moves/adds/changes can be eliminated, because their MPOA server automatically places hosts in their virtual subnetwork.

An additional use for virtual LANs is in the area of policy administration and security. Virtual LANs can divide the network into a group of hosts and can restrict the access these groups have with the servers. In this way, the virtual LAN acts as a firewall to provide additional security. Without virtual LANs, the network administrator needs to establish filters and access lists in routers or on servers. Device-centric or port-centric virtual LANs' configuration/management allows the network administrator to support filtering functions at the granularity of IP addresses, TCP port numbers, or MAC addresses.

Virtual LANs are typically limited only by the security of the managing database or the facilities the network layer protocols (e.g., TCP port numbers) provide by their support of MPOA servers. The major drawback to the current techniques used to build virtual LANs is that they are not standards-based and are potentially limited in how large the total system can become. MPOA, on the other hand, provides the functionality with standard-based technology and is not limited with regard to scalability.

Applying ATM to virtual LANs is attractive from both a manufacturer's and a network designer's perspective. Manufacturers view virtual LANs as an opportunity to focus their skills on building specialized networking hardware having fewer complexities than conventional multiprotocol routers. Ideally, the functions required to pass a packet from the source to the destination can be subdivided and assigned to different devices as shown in figure 12.1. For example, a LAN emulation bridge, which can be integrated into an MPOA network, is a product that only understands how to register itself on the ATM LAN and communicate with the ATM servers. With only these capabilities, and no knowledge of routing protocols, it can participate in a complicated enterprise network.

To a network designer, virtual LANs are attractive, because they provide the ability to invest in one common networking fabric for the entire enterprise network. For example, several ATM switches may be interconnected on a campus network to form the backbone. The attached devices may be a set of router servers, LANE bridges (media converters), and workstations. The network administrator, through virtual LAN technology, can use one physical network to interconnect all these devices; however, the workstations can be isolated from the router servers and the bridges will form boundaries. In this way, the network is centrally managed at the router servers, but is logically "firewalled" into many separate administrative sections by placing devices on separated virtual LANs or behind LANE bridges.

Segmenting the enterprise ATM network into multiple virtual LANs is also desirable to improve scalability. Dividing the network is useful so that broadcast and multicast domains are limited to a reasonable size, hence, the total network load from this traffic is localized. Limiting the distribution of broadcast traffic may be especially important in the LAN emulation model, where broadcasts may be more frequent. Scoping multicast traffic will play an increasingly important role, as bandwidth-hungry multimedia client/server applications become more prevalent and, as pointed out in chapter 11, reduce the number of virtual circuits concurrently carrying the same data.

12.3 MPOA's capabilities

The two techniques for building ATM networks discussed in chapter 11, LAN Emulation and Classical IP over ATM, can be used to build basic virtual LANs, because they permit the network to support multiple logical subnetworks on one switching fabric. In the ATM Forum's LAN Emulation model, a virtual LAN is equivalent to an emulated LAN. Interemulated LAN communication requires a router that is a member of two or more emulated LANs. In the Classical IP over ATM model, the philosophy is very similar. In this model, a virtual LAN is a group of hosts aggregated into a Logical IP Subnetwork, or LIS. The ARP server controls membership in the virtual LAN. As with LAN Emulation, intervirtual LAN communication is made by routers, which are members of multiple Classical IP Over ATM networks. When dealing with routable protocols, each of these virtual LAN models is applied to networks on subnetwork boundaries or domains.

Building upon these protocols the MPOA working group has gone through a lengthy process of developing a protocol that enhances LANE/CIP. The baseline capabilities of MPOA's virtual LANs empower the ATM network designer with several new capabilities for:

- Dividing the process of routing from switching
- Subdividing a network into logical workgroups on one common ATM network infrastructure
- Providing the potential to aggregate large amounts of traffic, on very high speed connections, towards servers that are members of many virtual LANs
- Providing the capability to act as a filtering or firewall device by restricting the traffic flows
- Providing centralized maintenance of workgroup membership, making moves, adds, and changes easier
- Providing improved performance for cut-through paths between ATM edge devices.

Finally, MPOA's virtual LANs empower the designer with a standards-based tool to build multivendor virtual LANs. It is important to note that the basic phase 1 MPOA specification is a document describing how to build networks using LANE bridges that know how to communicate across subnetwork boundaries. The bridges use the Next-Hop Resolution Protocol (NHRP) to determine the correct egress location on the ATM network.

12.4 Introduction to Multiprotocol over ATM architecture

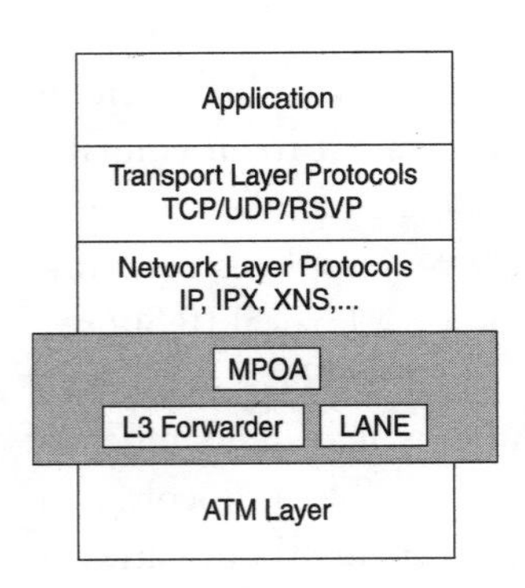

Figure 12.2 MPOA client protocol stack

In the Multiprotocol over ATM model, a virtual LAN is similar to a virtual subnetwork or virtual network. However, intervirtual LAN connections are not necessarily mediated by routers. With MPOA, hosts are capable of directly communicating with each other, even in the case where the path is between different Logical IP Subnetworks. (See figure 12.2).

In the MPOA model, the determination of the path data can take is based on decisions made by the edge device*. The traffic can be forwarded along the default path established between routers on the subnetworks, in the same manner as LANE or Classical IP. Or, a new virtual circuit can be created that interconnects the two dif-

---

* This discussion assumes best-effort service classes.

ferent subnetworks and passes the flow. This type of connection is known as a cut through or shortcut path. The important distinctions between these points are the decisions made by the edge devices and the measurement of flows.

Basic MPOA can take several different forms. First, a LANE bridge can provide the service by communicating with the MPOA servers that supply LES/BUS services. This will allow users who have implemented LANE products to redeploy them in an MPOA network with little or no modification. When the LANE bridge is upgraded, it can monitor the traffic leaving the subnetwork and know how to resolve the ATM address of the destination. It will then be able to support the basic features of MPOA.

Traffic monitoring by the edge device is important, because this supplies the edge device with a means to determine when the shortcut path should be created. Resolving the ATM address of the destination is critical once the edge device discovers the flow, because it must find the ATM destination of the target and establish a virtual circuit to the target. In some cases, the next-hop servers will reply with the address of the closest router to the egress of the ATM network nearest the destination computer. When the edge device can perform these functions, it allows the devices behind it on the Ethernet to benefit from these shortcuts, without their knowledge. The MPOA model operates by relying on multiprotocol route servers to maintain knowledge of the location of either the devices or the ATM network egress device closest to the ultimate destination. When the location is found, the ATM network can then place a call directly between hosts, thus relying on the low latency ATM network and eliminating the potentially high delay incurred when routing individual IP packets.

The MPOA server model is similar to using directory assistance on a voice network. The directory assistance operator is asked for the phone number of a destination and returns the value. With the phone number, a call can be placed to the desired destination. However, with MPOA there will be no restriction on local versus long-distance, directory assistance and the phone call may be placed using differing degrees of QoS. However, the QoS determination is left to the discretion of the end system. When a network designer considers deploying an MPOA network, there will be a choice of which protocol to use for resolving intrasubnetwork ATM addresses.

The designers of MPOA also faced this choice, and the conclusion was to use LANE as the core building block for intra-LAN address resolution. The choice of LAN Emulation versus Classical IP over ATM is primarily based on the types of traffic potentially carried on the network. If performance alone were the key factor and just IP packets would be carried, then Classical IP over ATM may be the best choice. However, if more than one network layer protocol is being implemented, then LANE is the only viable solution. LAN Emulation's great strength is that it enables network designers to treat ATM as a bridged technology for legacy intra-LAN communication. Classical IP over

ATM, on the other hand, treats ATM as a point-to-point link between hosts or routers, much the same as a WAN circuit.

An interesting side effect of choosing between Classical IP and LANE is the set of standards associated with the protocols. When building IP networks on ATM, the user can focus on just IEFT standards from the Integrated Services over non-broadcast multi-access (ION) network working group (NHRP and ARP servers). When working in a multiprotocol world, the set of standards used comes from the ATM Forum's MPOA.

12.4.1 MPOA benefits

In light of the preceding discussion, benefits of the MPOA solution can be characterized as follows:

- It provides the connectivity of a fully routed environment, where the router servers are few, but powerful devices, with large topographic databases. Thus, the MPOA solution separates routing and switching from routers and edge media converters taking advantage of ATM. It directs interdomain connection for best-effort traffic and, in some cases, QoS, providing a unified approach to layer 3 protocols over ATM via default forwarding followed by segregation postflow detection.
- It standardizes this new paradigm and the separation of routing decisions from data forwarding, which has been ongoing for several years in multiple standards bodies. Only recently have these forces come together to develop a common set of interfaces and protocols. The ATM Forum is a major focal point for this effort, and its work on new protocols has been allocated to the MPOA working group. This group is influenced by, and works closely with, the IETF's Internetworking over NBMA group and the Integrated Services group, which developed RSVP. These two working groups have been developing solutions to address the shortcomings and improve the functionality of LAN Emulation and IPv4 when run over ATM. The cooperative effort has yielded the Next-Hop Resolution Protocol, the Resource Reservation Protocol, and the MPOA document. These will form the basis of the next generation of networks.

12.4.2 Next-Hop Resolution Protocol

Consider this scenario: Every time two people want to hold a conversation, they have to use an interpreter or middleman. In this situation, delays in getting messages passed and increased complexity would be the norm. Ironically, this mediated conversation example is very similar to running applications over existing IP router networks, because the interconnection is always done via routers.

As discussed in chapter 11, both Classical IP over ATM and LAN Emulation suffer from the limitation that hosts in different logical subnetworks, but attached to the same ATM network, must also communicate via routers. The traditional model of routing is shown in figure 12.3. In order to improve upon this system, a new paradigm is needed that allows the hosts to communicate.

Directly establishing communication between hosts in different subnetworks, without using a router, is very desirable, because of the potential improvements in performance and scalability. This technique is referred to as cut-through or shortcut routing because it bypasses routers and cuts a path through the ATM network, or any nonbroadcast multiple access network, as illustrated in figure 12.4.

Cut-through allows hosts to find the ATM layer address of the ultimate IP destination or the router closest to the destination. Cut-through is also designed to work even if the destination is in a different administrative domain or subnetwork. Once the ATM address has been found, the idea is to circumvent the intermediate routers and establish a direct communication virtual circuit to that destination. Cut-through uses a device known as a Next-Hop Router Server (NHS) to acquire information about the host on the network, but it then relies on layer 2 technology, such as ATM or frame relay, to communicate with the remote host. The NHSs play a critical role in maintaining address information for the network, in much the same way as the directory assistance operator.

An important point to remember when discussing the implementation of NHRP is that its designers consider the technology to be one that is restricted to the topological

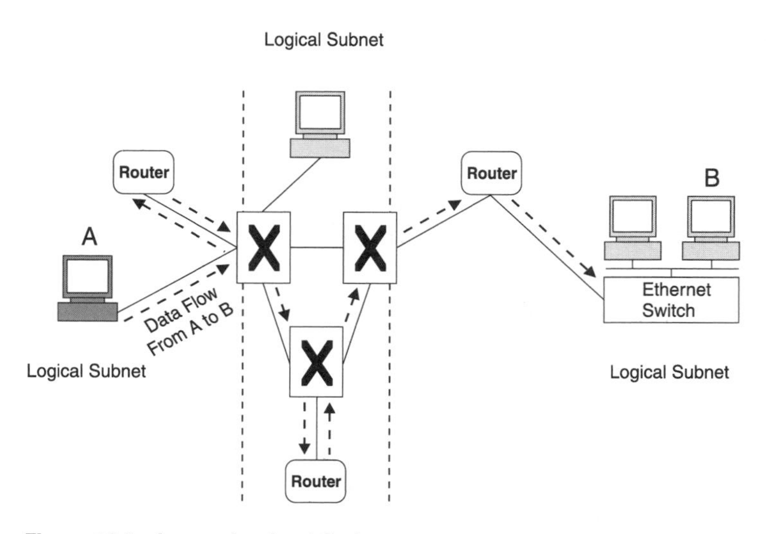

Figure 12.3 Large cloud network

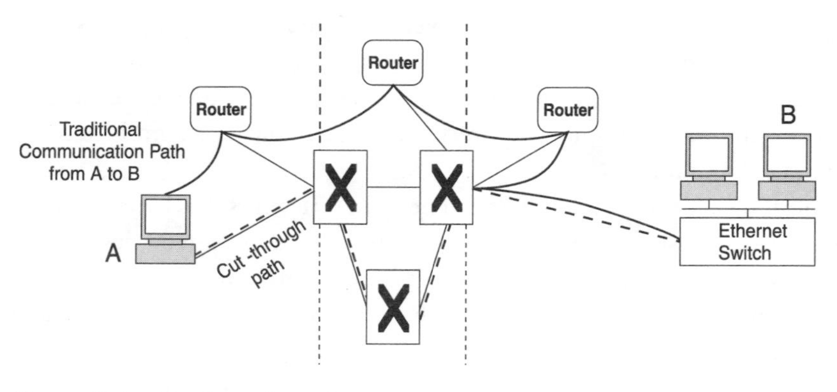

Figure 12.4 Cut-through routing

boundaries of an enterprise network. NHRP is an excellent technology for creating shortcut paths across an ATM campus network interconnecting routers, however, its ability to shortcut through a global ATM network is unknown. Therefore, the rollout of the service into next-generation networks will flow from the campus networks, next to possibly include ISP networks, and finally allowing cut-through of global ATM networks. The reason that NHRP is restricted to topological boundaries is that the process provided by the directory assistance does not scale well across administrative boundaries.

12.4.3 The "Large Cloud Problem"

Establishing a direct virtual circuit that crosses multiple administrative domains is a very complex problem, which is referred to as the "Large Cloud Problem." One can appreciate this problem by considering conventional routing protocols. Typically, routing protocols operate by summarizing or aggregating information to build their routing tables. When information is summarized, details about the layer 2 technology are either lost or hidden by the routing protocol. For applications to establish direct communication across an ATM network, they need the details of exact locations of destination, not just summarizations. MPOA/NHRP, in performing cut-through, also needs these details (e.g., ATM addresses) to set up virtual circuits between hosts.

To cope with this problem, MPOA/NHRP has an associated query protocol which can be used as a probe to follow the routed path to the destination. This protocol is capable of removing the aggregation of the route prefix and distilling the actual layer 2 address of the destination. Each query is generated by the edge device or the host and then passed along the default path towards the target. The query is passed, hop by hop among the router servers, until a server is reached that contains the mapping of the IP address to the ATM address. When the query protocol reaches this final router server it asks for the layer 2 address of the destination computer. When the ATM address is returned to the query generation host/edge device, the source can use it to establish an ATM switched virtual circuit that cuts through the ATM cloud.

NHRP is designed to allow end stations to locate each other via an "extended" ARP on networks, such as frame relay, X.25, and ATM. In a network supporting virtual circuits, devices attached to the same network must establish paths or calls in order to exchange data, but, unlike LANs, they lack the ability to easily broadcast a message to all hosts. In the NHRP model, these networks are known as "Nonbroadcast, Multiaccess" (NBMA) [1].

The NBMA Next-Hop Resolution Protocol allows a host or router to determine the internetworking layer addresses and NBMA addresses of the suitable NBMA next-hop towards a destination station. This address in effect will be that of the true destination or a device relatively close to the destination that will act as a data proxy. A subnetwork can be nonbroadcast and therefore can benefit from the shortcut routing, either because it technically doesn't support broadcast, as is the case with frame relay and ATM, or because broadcasting may not be feasible, as is the case with large SMDS networks. If the destination is connected to the NBMA subnetwork, then the NBMA next-hop becomes the destination station. Otherwise, the NBMA next-hop is the egress router from the NBMA subnetwork nearest to the destination station.

NHRP describes a next-hop resolution method that relaxes the forwarding restrictions of the LIS model. When the internetwork layer address is IP, once the NBMA next-hop has been resolved, the source may either immediately start sending IP packets to the destination, on a connectionless network, or may first establish a connection to the destination with the desired bandwidth and QoS characteristics, if the hosts are connected to a connection-oriented network.

When MPOA employs NHRP for shortcut resolution, a Next-Hop Server (NHS), which may be coresident with the LANE servers, provides the NHRP function of temporarily holding client addresses and responding to queries. The function of generating queries and establishing direct virtual circuits is the responsibility of the Next-Hop Client (NHC)* residing in the ATM host or ATM edge device. The NHS maintains a cache containing the network layer addresses (i.e., IP) to ATM address mappings. The cache is built either by having end nodes register at initialization time or by propagating the cache with values learned from the operation of NHRP over time. On the client, there is also a cache of address mappings that are learned from the operation of NHRP resolution reply messages or manual configuration.

---

* The NHC is sometimes referred to as a Multiprotocol Client (MPC). In this text, the two terms can be used interchangeably.

As with LANE components, before the NHRP process can begin, the NHCs and NHSs must be initialized. NHCs are initialized with the NBMA (ATM) address of their next-hop server and, of course, their network layer address. The NHS is configured with its own ATM address at startup, and it may be configured with the NHC's network layer address prefixes it serves.

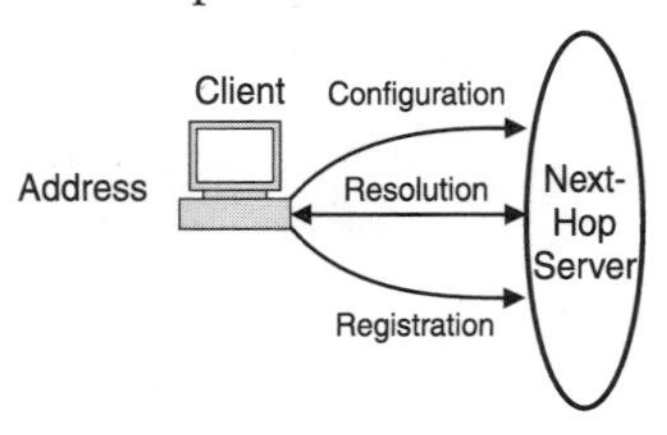

Figure 12.5 NHRP phases

When hosts use NHRP there are three distinct phases (see figure 12.5).

1. Configuration
2. Registration
3. Address resolution.

NHRP clients and servers participate in the phase by exchanging messages. In the configuration stage, the client must be configured with the ATM address of the NHS that is serving its domain; typically, this information is manually provided. The servers are configured with their own IP and ATM addresses. In addition, the server is configured to know which IP addresses are in its domain, that is, which addresses it takes responsibility for when NHRP queries arrive.

Subsequent to configuration, the clients register with their NHRP server and provide ATM addresses and IP addresses. When the clients have registered themselves with the server(s), the process of address resolution can begin. The process of resolving queries, from a high level of abstraction, is straightforward. The NHSs receive queries along the default IP route. If they do not maintain the domain, then the query is forwarded along the default path towards the NBMA destination, where the NHSs continue the process of address checking.

12.5 NHRP message types

Very similar to a client/server network, NHRP supports two basic message types: a query and, subsequently, a reply. In creating the shortcuts, the client first registers, and then the next-hop servers synchronize their clients' database information. Finally, the clients can issue queries. From the basic message types, the complete set of operations can be further expanded as follows:

1. NHRP registration request is used to explicitly register an NHC's NBMA information with an NHS.

2. NHRP registration reply is the reply issued by the NHS upon successfully registering a client.
3. NHRP resolution request asks for the network layer to ATM address mapping, which provides the necessary information prior to establishing the shortcut VC.
4. NHRP resolution reply returned by the NHS having responsibility for the target and contains the NBMA address.
5. NHRP purge request explicitly requests the deletion of an NHC cache entry.
6. NHRP purge reply acknowledges the deletion of the NHC's address from the NHS's cache.
7. NHRP error indication signals an error condition, for example, unrecognized extension, protocol error, or invalid reply received (see table 12.1).

Table 12.1 NHRP Error Codes

Error Code	Description
1	Unrecognized extension
2	Subnetwork ID mismatch
3	NHRP loop detected
6	Protocol address unreachable
7	Protocol error
8	NHRP SDU size exceeded
9	Invalid extension
10	Invalid Next-Hop Resolution reply received
11	Authentication failure

12.6 NHRP operation

The edge device or ATM host proceeds in a client/server-like manner to initiate the operation of the protocol when a source (S) has data to transmit to a destination (D). This can be caused by flow detection in the edge device or by an explicit shortcut creation generated by the ATM host. The first step in NHRP is the creation of an NHRP resolution request packet which is transmitted towards the destination along the default routing path.* The resolution request contains the source's network layer and ATM address, along with the destination's network layer address. If the source has additional

* In practice, the NHC's default router is typically also its NHS.

data to immediately transmit to the destination, as may be likely with an edge device, it can continue to transmit the packets along the default path until the shortcut response is returned and the short-cut VC is created. (See figure 12.6.)

The second phase of NHRP is the processing work done by the NHSs to find the ATM address belonging to the target. As each NHS receives the resolution request packets, it checks to determine if it is responsible for the target's IP-ATM mapping. If it does not serve the destination, it forwards the packet to the downstream NHS. An important detail, critical to the protocol's operation, is that the downstream NHRP servers must not generate NHRP resolution requests for data they receive over their ATM interfaces. If they did, it would cause each downstream router to generate an NHRP request for the data flow. Therefore, every router along the default path would request a virtual circuit to the destination. This is called a "domino" effect and is clearly undesirable.

As the NHRP resolution request traverses the network, the NHSs run address cache checks. If the present NHS does maintain the target's LIS, then it replies to the source with the target's ATM address. Also, if the path to the target is through a router

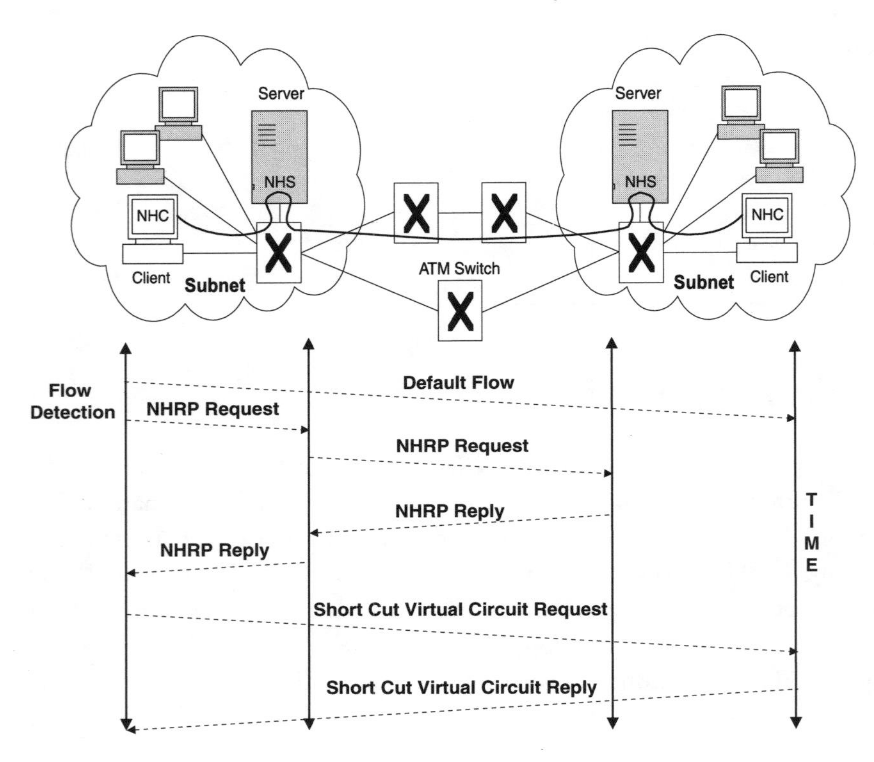

Figure 12.6 NHRP message flow

or bridge placed at the egress of the ATM network, then that device's address is returned in the NHRP reply. If no NHS on the ATM network possesses an IP-to-ATM address mapping matching the NHRP Request, then a negative (NAK) NHRP resolution reply is returned.

The NHRP reply can contain a single IP/ATM mapping or an aggregate mapping of several IP addresses to one ATM address. The latter would be useful in cases where the egress to the ATM network is a bridge. In this case, the bridge would generate an NHRP reply containing the target's address information and a prefix length, which would, in essence, specify the amount and the portion of the IP address space accessible via the egress router.

As the NHRP reply is passed along the default path back to the source, the NHSs along the path have the ability to read the response and locally cache the result. When the result is cached, the local server can reply to subsequent NHRP requests for a known destination. If this behavior is undesirable or not trusted, then there are two mechanisms that can be used to turn this feature off. First, the source can request an "authoritative" reply, which means that only the NHS that truly maintains the address mapping can reply. Second, the cache entries can be systematically purged, which will force the NHRP request to travel to the destination's NHS.

12.6.1 Address aggregation

There are some implications in address aggregation and egress responses that must be considered in order for NHRP to operate correctly. In order to successfully receive replies to inquires, all egress routers should support the protocol and reply with NHRP requests for the subnetworks they serve. If the NHRP requests have reached an egress router along the default path, it means that the IP layer routing protocols will forward best-effort traffic to this device and, therefore, should also accept shortcut VC creation. Two problems with address prefixes may be encountered:

- In the worst case, the egress router will not reply to the NHRP resolution request and the default forwarding path will be used for the duration of the data.
- Additionally, if the egress router advertises too large a prefix, a "black hole" will be created in the network.

If the router fails to support address aggregation and serves as an egress for a large section of the intranet it will, most likely, be the recipient of several unnecessary NHRP requests. Finally, because the NHRP messages are trying to resolve a mapping between a network layer address and an NBMA ATM address, there is no reason for NHRP queries to leave the ATM network. Therefore, at this point in the standardization process,

NHRP messages should terminate at an egress router when the subsequent path is not an NBMA network.

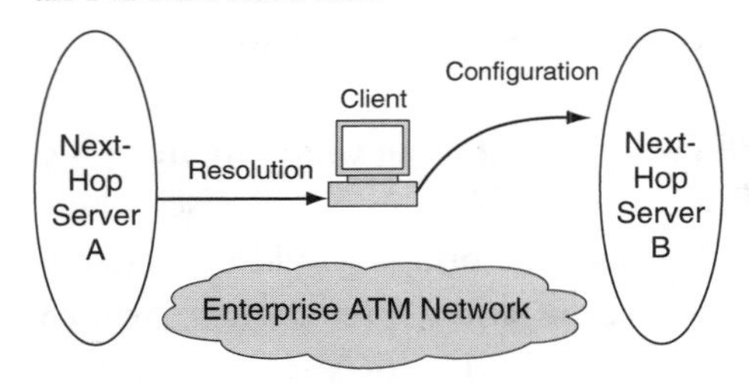

Figure 12.7 Multiple NHSs

An additional point of concern involves the case where the LIS is served by more than one NHS/router as shown in figure 12.7. In this case, the client may have registered with only one of the servers. If an NHRP resolution request were to arrive at the other server, then there is the possibility that a negative acknowledgment would be returned. To avoid this situation, the servers exchange their state using a database synchronization protocol known as SCSP. With SCSP, when a client has registered with one of the servers, it does not need to register with the secondary NHS in order to have its resolution request correctly processed.

When an NHS determines that it serves the target, there are several ways that NHRP replies can be returned to the requesting host. First, the basic mechanism is to have the NHRP reply follow the default routing path back to the source. If this path is chosen, it has the benefit of allowing the NHSs along the route to update their caches with the newly acquired IP/ATM address mapping. Upon receiving subsequent NHRP requests for the same target, the intermediate NHSs can reply directly from their caches. NHRP allows authoritative and nonauthoritative formats. If a host wishes to get locally cached information, it can issue the latter message type. If that message fails, an authoritative request is generated.

A second technique for returning the NHRP reply is to forward it directly to the requesting machine, bypassing all intermediate NHSs. This is possible in networks where the ultimate NHS can reach the source either via a direct VC creation or on a connectionless medium, such as SMDS. The benefit of replying directly to the NHRP requesting machine is that it can reduce the latency encountered when the reply is returned along a long string of NHSs.

12.7 NHRP extensions

NHRP has been designed to provide for expansion to include features that may not have been thought of by the protocol's designers. It allows the query to contain one or more additional messages, called extensions, within the NHRP query (see table 12.2). These are used to carry information that is outside the core scope of next-hop address resolution. Some possible uses for extensions are as follows:

- The NHC client may request, via an NHRP resolution request extension, the ATM address of the NHS generating the reply.
- Clients can request the addresses of all the NHSs along the path between them and the NHS that generated the reply.
- The client or servers may authenticate each other's messages with an authentication extension.
- Finally, the NHC and NHS can pass information only usable to a certain vendor specific application via the vendor private extension. This may be used to help build a special signaling message or to modify signaling parameters.

Joining three fields, type, length, and value (TLV), in a string forms extensions to NHRP packets.

Table 12.2 Sample Extensions

Extension	Description
NHRP Forward Transit	Lists transit NHSs along path from source to destination
NHRP Reverse Transit	Lists NHSs whose reply crosses from destination to source
NHRP Authentication	Lists authentication information used to validate query
Destination prefix length	Indicates that destination can reach a set of IP addresses (i.e., the destination is either multihomed or an edge device)

12.8 Distributed routing

Prior to examining the MPOA architecture in detail, the last topic requiring further explanation is the concept of a distributed router. Distributed routing is the realization of separating the higher-level functions of route determination from the lower-level functions of switching data before, or during, the time data pass through a network. A distributed router consists of a central router server, which controls multiple edge devices, as shown in figure 12.8. In the distributed router model, the edge devices do most, if not all, of the data link layer switching.

Together, the route server and edge devices are used to build a layer 3 protocol independent of distributed architectures and also exhibiting independence in physical implementation. Routing servers in this model run traditional legacy routing protocols along with ATM's routing protocols (i.e., PNNI) and supply routes to the edge devices. By isolating the router server function, it is possible to maximize the use of a very expensive resource: the routing engine.

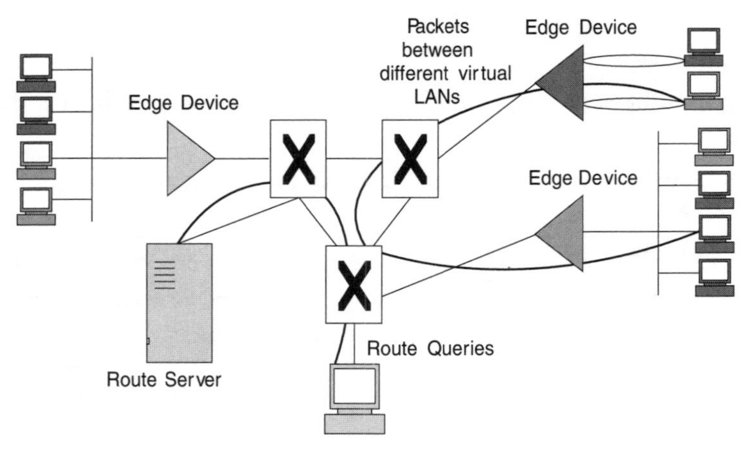

Figure 12.8 Distributed routing

Route servers can be constructed with a very high-performance, computationally powerful machine that can serve an entire campus. In this model, MPOA edge devices are seen as simple bridges detecting flows, generating NHRP messages, and then establishing direct virtual circuits. This distributed router is an excellent way of conceptualizing MPOA. In MPOA, functions are defined and the actual implementation of these functions is distributed throughout the cloud.

12.9 Multiprotocol over ATM requirements

The ATM Forum, in close cooperation with the Internet Engineering Task Force, established the goals and model framework of MPOA in late 1995. Since then, the ATM Forum working group has seen the protocol through several stages of evolution. The original intent was to generate a document that resolved all the problems of shortcut path selection, QoS routing, multiprotocol operation, and so forth. However, it was determined in late 1996 that achieving the original goal would not be possible if the specification was to be completed by the end of the year. Therefore, the phases were agreed upon, with shortcut and legacy interaction being the scope of the initial requirement set.

Two of the principal goals of MPOA's first phase are allowing different administrative domains to concurrently exist on one physical ATM network and supporting communication between any two devices running layer 3 protocols.

MPOA leverages much of the knowledge gained by the ION working group and extends the model to allow multiple protocols and distributed routing architectures.

This will be accomplished in the realm of lower-cost edge devices which are network-layer intelligent due to extracting router server functions. The remainder of this chapter focuses on the MPOA model. Using QoS in the first phase of MPOA is then relegated to RSVP over ATM, as described in the previous chapter.

When the protocol development is divided into different phases, MPOA's critical phase 1 requirement can be further described as follows:

- Allowing MPOA devices to establish direct ATM connections using NHRP
- Integrating with LAN Emulation in the edge device and allowing the MPOA server to support LANE's servers
- Providing support for firewalls and protocol filtering by allowing the network administrator to create virtual LANs on the layer 2 fabric
- Supplying support for broadcast traffic
- Providing support for automatic configuration of ATM hosts via an initialization protocol
- Illustrating the concepts of separating switching and routing concerns
- Multicasting within the subnetwork.

Subsequent releases of the MPOA document are still having their scope changed; however, the more likely alternatives that will be addressed are the following:

- Multicast using the Multicast Address Resolution Server (MARS) as described chapter 11
- QoS, as described in chapter 10, using RSVP [2].

As can be expected, the product of the joint effort between the ATM Forum and the IETF has generated substantial interest, both positive and negative, within the industry. The completed specification reduces limitations found in previous network models and allows customers to work in a multivendor environment. In addition, it will provide a higher-performance and more scalable solution for multiprotocol LAN internetworking over ATM that will enable next-generation intranets/extranets.

12.10 Multiprotocol over ATM operation, components, and architecture

An MPOA network consists of several network layer-aware components which can be subdivided into router servers, edge connection devices, LANE servers, next-hop servers, and ATM-attached hosts. In the MPOA model, the ATM fabric is considered to be one physical network capable of supporting many virtual LANs. Each network, while separate, is reachable using shortcuts. In a sense, the ATM network can be visualized as an emulated multiprotocol bridge/router with the addition of very high bandwidth capabilities (e.g., ATM with MPOA yields a virtual router). The network is known as a virtual router because with MPOA the process of routing is separated. (See figure 12.9.)

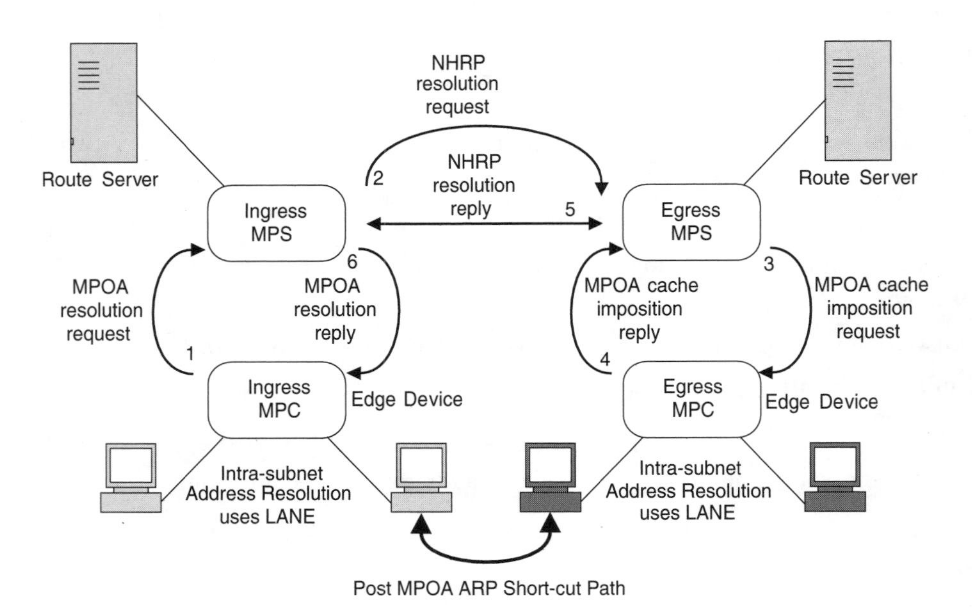

Figure 12.9 MPOA components

An MPOA network is built from special-purpose address resolution servers, edge bridging devices, and new software on ATM attached devices, such as hosts, that allow it to commute with the server. The actual protocol documented in the MPOA specification describes which pieces are used to build the network, along with the information flows among these components. The low-level details of implementation and separation of actual processes running on devices is left to the manufacturer or network designer. The MPOA model only defines logical components, message flows, and suggested behavior, not product.

In the first release of the MPOA specification, the key architectural components are as follows:

- Edge connection devices are used to physically attach legacy networks to an MPOA system (i.e., an Ethernet to a ATM converter). These are similar to LANE bridges. An MPOA client is only responsible for maintaining a listing of layer 2 addresses to layer 3 for the host(s) with which it is communicating.
- Route servers have topological information gained by running routing protocols and distributing state among themselves. These are network components that support MPS functions along with LANE and NHRP. An MPOA server maintains knowledge of layer 3 protocols and topologies for the areas served.
- Information flows comprise the protocol descriptions for MPOA Client (MPC) to MPOA Server (MPS), MPC to MPC, and MPS to MPS exchanges.

12.10.1 Information flows

An MPOA system utilizes several information flows. The information flows describe how the components exchange MPOA state information and resolve target addresses and state. The MPOA system works by allocating tasks to groups and then defining the protocol's operation by specifying information flows among the groups.

In defining the information flows between functional groups, the ATM Forum has broken the problem into specific cases corresponding to different states of the protocol. From a high level of granularity, MPOA's logical components can be divided into clients and servers, so that the implementation of the protocol follows four distinct steps.

1. Configuration is used to retrieve configuration information and register it in the network.
2. Discovery is the phase where the MPOA devices learn their topological relationships.
3. Address resolution flows are used after configuration to query the MPOA servers and then inform the servers of state changes on the client side. It is also an informational flow between MPOA servers containing routing information.
4. Data transfer is the actual transmission of information between hosts using an MPOA system.

These protocol phases can then be further subdivided into information subflows that perform the detailed operations. For example, client members of an MPOA system pass information to the following:

- Their route server to obtain address resolution information
- Their LANE LES for destination resolution of intra-LAN traffic

- Their LANE BUS, which will forward broadcast data in the absence of a direct point-to-point virtual circuit.

MPOA servers pass information to establish peering conversation with other MPSs and to communicate with edge devices regarding topology information.

12.10.2 Configuration

Before the MPC and MPS can begin using the MPOA system, they must be registered and configured. The configuration process is accomplished either manually by a network administrator or the MPC/MPS can make use of a LANE configuration server as described in chapter 11. As in LANE, devices in an MPOA network usually contact the configuration server at boot time. The configuration server knows which clients and servers are associated within which virtual networks, and the configuration server notifies the clients and servers of their respective MCS/MPS ATM addresses.

When route servers are initialized, they pass a TLV, identifying themselves to the LECS specifying a configuration request. They are given the identity of the subnetwork(s) they control, along with the layer 3 protocol type(s) used. In addition, the route server is a member of the subnetwork(s), so it also acquires a layer 3 address.

The MPCs then register with the LECS by sending a configuration request containing a TLV identifying the MPC. When MPCs and ATM-attached hosts are initialized on an MPOA network via the LECS, they are given information about which policies should be followed for shortcut setup, when to time out and delete idle virtual circuits, and which protocol (e.g., IP or IPX), should be using short-cuts. In addition, the ATM address of the MPS can be forwarded, however, this information could be inferred to be the default router's IP address. The mapping from Network Layer address to ATM address on the default router would be learned via LANE.

12.10.3 Discovery

The discovery phase occurs after initialization and concerns the set of information exchanges used to inform each MPOA device of its existence, capabilities, and domains. The term, "discovery," when related to MPC, describes the ability to determine the location of the NHS. As with the configuration phase, the mapping from network layer address to ATM address is done via LANE LE_ARP. Once the discovery phase is complete, the MPOA components within a domain can pass NHRP/LANE messages, and they can now begin to allow MPC to communicate among themselves across subnetwork boundaries.

12.10.4 Address Resolution

Once a host has been configured and has registered itself on the network, it can begin to communicate with other hosts. In order to communicate, the mapping of layer 2 to layer 3 address must be resolved via the MPS. Data flow between computers on an MPOA system can be one of three types:

1. Intrasubnetwork via the LANE (or Classical IP) servers
2. Intersubnetwork via default forwarding
3. Intersubnetwork via shortcut routing.

Intrasubnetwork communication takes place using the mechanisms of primarily LANE but in some cases can be accomplished via Classical IP. The choice between the two is based on the expected internetwork layer protocols used on the network. When the MPC connected to non-NBMA receives a flow of packets, it attempts to determine, by examining the internetwork address, if the flow should be short-cut via NHRP.

The multiprotocol servers along the path will generate and process NHRP messages in the manner described previously. Each NHRP message is forwarded, hop by hop, until it reaches the egress MPS. At the egress MPS, a message is passed from the server to the MPC to determine the ATM parameters for encapsulation as described in RFC 1483 [3]. The MPC serving the legacy-attached host replies with addressing information to the MPS, so that it can successfully generate an NHRP reply. Before the MPC replies to the MPS, it must, however, ensure that it has sufficient resources to accept a new virtual circuit. At that point, the egress MPS can generate an NHRP reply and return it along the path the resolution message took.

12.10.5 Data Transfer

When an MPC has successfully received a response to its NHRP query, it can establish a direct virtual circuit and begin transferring user data. The parameters, such as PCR/SCR/MBR, used during creation of the ATM virtual circuits between MPC to MPC or MPC to ATM end stations are to a large degree left up to the discretion of the end user or the network administrator. However, several base rules have been specified to help ensure smooth default operation. MPOA specifies that, as a baseline, the parameters documented in RFC 1755 covering signaling parameters for Classical IP over ATM should be used for user data communication.

When selecting a service category, the default choice is Unspecified Bit Rate (UBR). If an MPC wishes to create a virtual circuit with another MPC that has a service category other than UBR, it can signal this wish via an NHRP extension. If the receiving MPC is able to support the source's QoS request, it can signal this fact to the source

MPC by returning the NHRP extension. QoS is left as an option for the end devices and should be negotiated with the RSVP protocol.

Virtual circuits used for data communication or to pass control messages can be deleted when their usefulness is no longer apparent. Most likely the deletion will be done by the edge device that has created the circuit after an idle period has been exceeded. For example, control circuits are deleted by default after 20 minutes of idleness. Virtual circuits between MPCs are deleted shortly after the traffic flow has stopped.

12.11 Summary of MPOA

This chapter has presented the MPOA protocol, as developed primarily by the ATM Forum with contributions from the IETF. A great number of the components required to build an MPOA system are intentionally based on architectures already in place in networks that support LANE or Classical IP over ATM. The key technologies involved in building a first-generation MPOA network are the Next-Hop Resolution Protocol and LAN Emulation, which are provided by the distinct functions of the MPS and the MPC, respectively.

Baseline MPOA systems can be built with edge devices and traditional IP routers acting as MPS or LANE servers upgraded to support NHRP. The current version of MPOA can be used to take advantage of ATM's ability to provide higher speeds and lower latencies when compared with traditional IP routers. The baseline of the protocol is currently being used to build products for early adopters and has been deployed successfully.

Subsequent versions of the MPOA standard will continue to bring into scope work on multicast and QoS. By allowing the standards bodies to develop protocols such as MARS and RSVP, which are very good at solving their respective problems, the MPOA group's work has become one of documenting protocol interactions and, when necessary, extending previous work to support ATM's special requirements and features. The current state of the protocol has yielded a completed document which has been used by vendors to construct compliant hardware and/or network designs based on the new hardware. The following chapter will describe how these products can be put to use and what alternatives to MPOA exist.

12.12 References

1 D. Katz, D. Piscitello and B. Cole, *NBMA Next Hop Resolution Protocol (NHRP)*, Located at: *ftp://ietf.cnri.reston.va.us/internet-drafts/draft-ietf-rolc-nhrp-07.txt*. December 29, 1995.

2 R. Braden, L. Zhang, and S. Berson, *Resource ReSerVation Protocol (RSVP)—Version 1 Functional Specification*, 02/22/1996. Located at: *ftp://ietf.cnri.reston.va.us/internet-drafts/draft-ietf-rsvp-spec-10.txt.*

3 J. Heinanen, "Multiprotocol Encapsulation over ATM Adaptation Layer 5," RFC 1483, July 1993.

CHAPTER 13

MPOA Migration Strategies and Alternatives

The preceding chapters have presented the foundation for building multiprotocol networks and how they are implemented. The key technologies revolve around RSVP, for resource reservations, NHRP coupled with PNNI for path determination, LANE for address resolution, and MPOA as an all-encompassing standards document. This chapter presents ideas for the migration of a network from a legacy model of router and traditional broadcast media towards one that implements some, and possibly all, of the components listed above. In addition, the chapter will discuss some of the potential pitfalls associated with the technologies and what competition exists to prey on these shortcomings.

To briefly review, the goals of MPOA are to bring together several different internetworking technologies to achieve a level of functionality that to date is unparalleled when each is employed independently. MPOA provides network designers with the capability of building systems that will accomplish the following:

- Allow multiple virtual LANs on one physical network and control access between the VLAN with router servers
- Empower applications with the ability to use QoS parameters as a standardized means via RSVP
- Allow clients to directly communicate with servers in different domains, without passing all data through routers via next hop servers
- Describe how to separate the functionality of routing from data forwarding by creating two distinct devices, router servers and edge devices, and specify a client/server model of inter-LAN route determination.

As described, the LANE and Classical IP over ATM specifications are effective in allowing users to quickly deploy ATM. However, if these technologies are used to speed deployment of ATM, then the user must compromise by giving up ATM's rich QoS capabilities. In addition, these technologies suffer from the same scalability problem seen in traditional networks, because inter-LAN traffic must be passed through routers. Exposing the ATM layer and allowing applications to make use of QoS parameters, while supporting virtual LANs, is a major benefit of MPOA.

Initially, MPOA and LANE networks were very similar and began to differ only when their size grew, or when users executed applications that passed data outside the LAN. Because MPOA and LANE are so tightly coupled, it seems reasonable for corporate planners to deploy LANE today with the intent on migrating to MPOA. As the network grows, administrators will be able to create additional virtual workgroups which can all share one ATM switch fabric. The drawback to this approach is the features lost by not using Classical IP. If the network is migrated from LANE to MPOA then the initial system will not provide good support for IP multicast traffic and will not support RFC 1577's larger packet sizes [1].

Regardless of whether the network is being migrated from LANE or Classical IP, one of the key issues addressed by the MPOA working group is interoperability with legacy routed architectures. The MPOA model of a virtual router composed of edge devices, router servers, and the ATM infrastructure is designed to work with existing routers using legacy router protocols such as RIP and OSPF. Therefore, network managers can use MPOA in conjunction with, rather than as a replacement for, their current routed networks. Because of the control offered in the virtual router model, MPOA implementations can be created in small test pockets.

13.1 Migrating to MPOA

Clearly, virtual LANs, using LAN Emulation or MPOA, are extremely useful tools for building multiprotocol corporate networks. The challenge now is to devise methods to successfully develop and deploy technology to meet the current and future requirements without causing undue service interruptions. MPOA is a first step in meeting that challenge as is evident when comparing the MPOA model to LAN Emulation. Consider the following points:

- MPOA is an evolution of the LAN Emulation model; therefore, MPOA will make use of LAN Emulation services.
- LAN Emulation operates at OSI layer 2 [2], hence, it is bridging.
- MPOA operates at both OSI layer 2 and layer 3, hence, it is both bridging and routing.
- LAN Emulation requires no modification to host protocol stacks; MPOA requires modification.

Corporate implementations of MPOA should be viewed as the combination of three tasks. First, network managers will need to understand the various components, protocol flows, and interactions with legacy systems. Second, they need to work with vendors to build a plan for the system infrastructure that can be introduced without disrupting current communication. Finally, they must devise a scheme to implement their architecture plan with a phased introduction of the technology into their intranet.

Migration to MPOA involves adding MPOA servers to the network and may be as simple as upgrading the ATM switch control software if the feature is supported, or, more likely, running the MPOA servers on ATM-attached routers. In order to connect legacy networks to the ATM/MPOA network, the next hardware component required

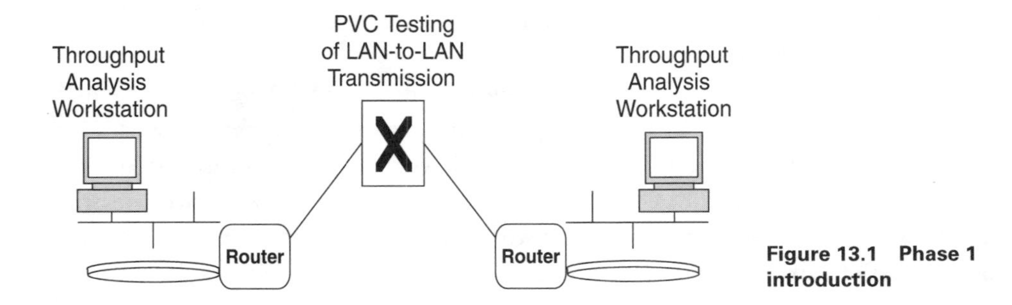

Figure 13.1 Phase 1 introduction

would be the addition of MPOA edge devices. In all likelihood, these will be achievable via software upgrades to LANE bridges. Finally, the ATM-attached host will need to have its LANE or Classical IP software updated to support NHRP, flow detection, and, optionally, RSVP.

Because the components are well defined and can stand alone, the introduction can proceed as follows with minimal risk.

- Introduce ATM components in a test environment to gain experience with their operations and network management. It may be necessary to maintain a long-term test network for verification of new hardware, such as ATM video servers or next-generation switches, before introduction on an enterprise network. Before deploying ATM technology, the network manager should feel comfortable with interconnecting legacy networks using Permanent Virtual Circuits (PVCs) and/or LANE servers, as shown in figure 13.1.
- Migrate test environment to edge subnetwork to gain experience with production traffic over an ATM/LANE routed network having a single path to the enterprise network over a non-ATM connection, as shown in figure 13.2

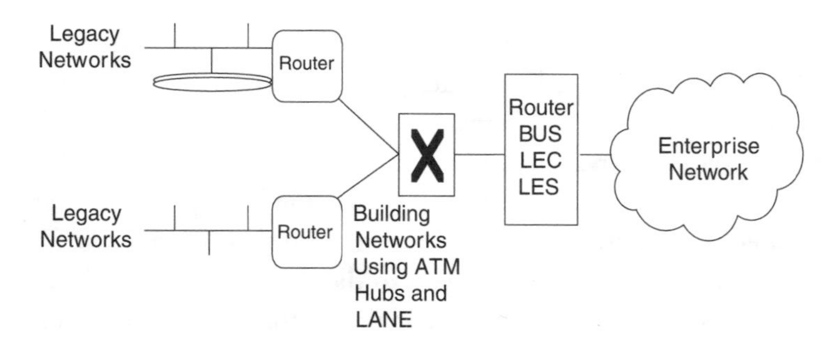

Figure 13.2 Phase 2 introduction

- Increase the number of ATM switches on enterprise network and gain additional network management. Begin interconnecting disperate ATM subnetworks and enable next-hop service.

If the MPOA network is phased in slowly, network managers can begin with a simple architecture of an ATM switch interconnecting routers via permanent virtual circuits. From that foundation, a router can be upgraded to support multiprotocol services and the other routers can be upgraded to multiprotocol clients. This will be, in essence, the first step in creating a virtual router by offloading function to a central location. As the routing function becomes more virtual and the network grows, the devices implementing the virtual router can become more specialized. For example, the second-tier routers are really edge devices.

In this migration plan, the edge devices and ATM switches become the network's workhorses, focused on high-performance data transfer and low-latency switching, and the router servers become the network's brains. In this basic migration scheme, traditional routers do not disappear. Instead, they either operate completely oblivious of MPOA or they migrate towards providing either MPOA services, termination of WAN connections, or traditional firewall filtering. By using this model, it seems most likely that ATM's role will be to deliver low-latency, cut-through operation on the campus backbone. When the same technology reaches the enterprise hub or desktop, its role is not clear; the battle will be between "simpler" technologies such as 100 Gigabit Ethernet.

The ultimate desktop technology battle may be won by the technology that does not necessarily provide all the bells and whistles of MPOA, but provides an effective combination of some features at lowest cost. Regardless of the migration path chosen or the desktop technology, the basic components of the MPOA solution will be routers, most of which are already in place, ATM switches, and ATM hosts. A major benefit of

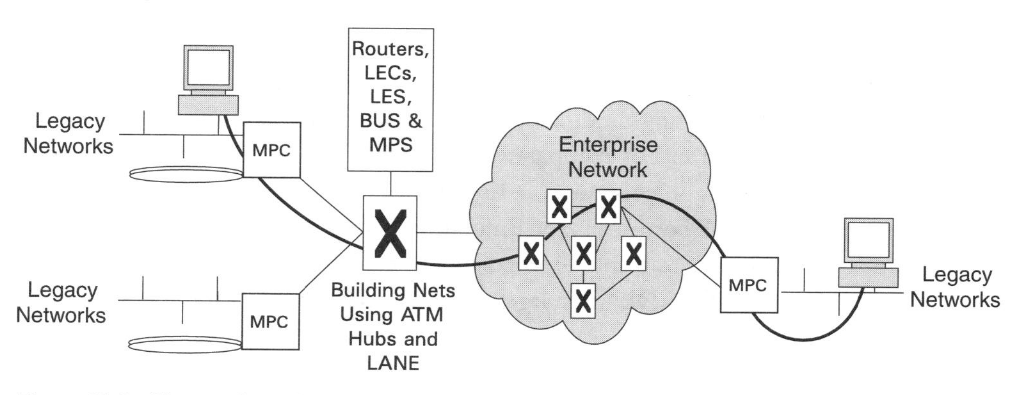

Figure 13.3 Phase 3 introduction

MPOA is the ability of the core network switches to be supplied by any vendor, because they are only providing a generic switched virtual circuit service.

Analysis of ATM network architectures and protocols is important; however, more importantly, network managers are faced with plotting a migration strategy from legacy LAN infrastructures to ATM-based LAN internetworks. As described, when migrating to ATM LANs, users will typically start with a pilot network based on ATM, and then expand as this network proves to be stable. When demand for bandwidth and enhanced management capabilities grows, the pilot network will need to be expanded into a production system. Subsequent to the stability of the ATM LANE, the network manager can then migrate to support MPOA services.

In pursuing such a plan of phase introduction, one has to take into consideration the following.

1. Plan for interconnecting the legacy network to the new ATM fabric.
2. Select a baseline model for intra-LAN address resolution (LANE or RFC1577), based on the internetwork layer protocol being used.
3. Plan for upgrading the pilot ATM network in multiple phases from both a feature and topology standpoint to MPOA or a cell-switching router technology. This will involve one of more of the following:
    - Addition of MARS servers for IP multicast support
    - Further growth of the campus network into the ATM fabric
    - Integration with the campus network management system
    - Addition of next-hop resolution servers and upgrades to LANE bridges.
4. Understand vendors' long-term plans for product roll out and upgrade strategy.

First, one needs to understand the various models outlined above and decide which is appropriate for a particular environment. Performance, features, and product availability have to be considered. For IP-only networks looking for good performance, the Classical IP over ATM is a strong candidate, because of the potentially faster ARP mechanisms and larger packet size. For companies that have a campus network supporting multiple internetworking protocols, such as IPX or AppleTalk, and where performance is not as critical, the ATM Forum's LAN Emulation model is the best starting point. Finally, companies that have multiple protocols and need to consider performance and scalability should start with LANE and target a migration to the multiprotocol over the ATM model.

Key components to assist in the migration from today's LAN internetworks to the newer architectures are ATM edge devices with a variety of legacy LAN interfaces

(FDDI, Token Ring, Ethernet) and routers with native ATM support. Both of these devices allow existing LAN internetworks to interconnect with new ATM-based LAN internetworking solutions. With the LAN Emulation model, one can link to legacy LANs through a LAN-to-ATM bridge or through a router with an ATM interface that understands LAN Emulation. With the Classical IP over ATM model, one must use an IP router to link legacy LAN internetworks to the Classical IP hosts. In the multiprotocol model, one can link to legacy LANs through a LAN-to-ATM network layer intelligent switch or through a traditional router with an ATM interface that understands the Multiprotocol over ATM protocol.

It is important to note that Classical IP over ATM and LAN Emulation are not mutually exclusive, and network planners may choose to have ATM-attached hosts running both Classical IP and LAN Emulation simultaneously. The only caveat is that the different models will not directly interoperate without intervention of a gateway device. An ATM host speaking "Classical IP over ATM" will not be able to communicate to a host speaking "ATM Forum LAN Emulation" without going through an IP router that understands both these models.

A host in a LAN Emulation network communicating with a host in the Classical IP over ATM network will have to go through an edge device or router using both these protocols. This approach may be useful for determining what the values of overhead and latency are with ARPing. Another reason for supporting both protocols would be if the enterprise network were a large IP over ATM cloud, but also contained pockets of other internetwork layer protocols that needed to ride over ATM without using IP tunneling.

Finally, network managers should strive to get guarantees from their vendors that the products they are buying can be upgraded to support the MPOA functionality planned for the long term. A user, for example, may start by implementing MAC-layer LAN Emulation today, but have plans to deploy Multiprotocol over ATM as the standard becomes finalized. If the products users buy today are not designed to accommodate additional network layer intelligence from a router server, a significant upgrade may be needed some time in the future. While initially the ATM networks are likely to be fairly simple, customers need to understand their vendors' rollout plans for new features and their own long-term architecture before making significant investments.

The completion of the first stable version of the MPOA specification was produced in early 1997, followed by a short lag time until standards-based products were available. Implementations of standards-based MPOA products on a corporate network is something that early adopters are able to do; however, most cautious network administrators will wait until the technology has been proven. The key issue network managers are dealing with on first phase networks is trying to understand the resources required to maintain the system. Examples of where these resources are unknown would be the number of required call setup messages processed per second, the total number of SVC that needs to be maintained, or the physical truck speeds required to interconnect next-hop servers.

13.2 *Alternative to MPOA: network layer switching*

During the development of the MPOA protocol, some vendors began to express concerns that the work being generated by the ATM Forum was too large and complex. There was a sense that ATM's growth had become too complex to be practical and implementable. They felt that in order to support multiprotocol networks, using ATM Forum technology, a vendor would need to support at least UNI 3.1 signaling, PNNI Routing, NHRP for shortcut resolution, LANE for address resolution, MARS for multicast, and so forth.

Of the problems introduced by large layer 2 networks, some of the more pressing concerns have to do with the ability of layer 2 switching to smoothly integrate with layer 3 switching. As mentioned, there are many vocal opponents of current protocols who claim that the ATM Forum and the IETF have taken radically different approaches. Also, scalability and ease of operation are critical when building large intranets and this combination is hard to find with a pure layer 2 switched network. High degrees of scalability can be difficult to achieve with a layer 2 switched network, because the address space is nonheretical.

In addition, standards activity and recent network deployments have raised issues surrounding traffic engineering, that is, the ability to use the layer 2 network to explicitly define the route data follow. Traffic engineering is one of the more pressing needs for ISPs and continues to drive much of the work described in this section. It is a problem that needs to be resolved in order for the Internet to realize continued growth.

This section will present protocols developed to facilitate the integration of layer 2 switching with layer 3 routing. These protocols produce efficient networking capable of being scaled up to very large internetworks possessing both efficiency and solid traffic management capabilities. This information is important for network designers and managers, because it illustrates the trend in intra/internet design, which is towards networks incorporating hybrid architectures comprised of many cooperating best-of-breed technologies. Network layer switching technologies represent this next-generation class of protocols, because, as will be shown, they are made up of the finest points from the data link layer integrated with robust and proven network layer routing protocols.

13.2.1 Motivations for network layer switching

Before beginning a discussion of network layer switching protocols, it would be helpful to review the pros and cons of data switching at the various layers of the OSI protocol stack. As has been shown in the previous chapters, ATM and other layer 2 switching technologies have been designed to provide very high speeds and low latency. This makes them good choices for delay-sensitive applications, such as voice/video data, or in cases where high bandwidth is critical, such as ISP backbones.

Layer 3 technologies, such as IP, have been in existence since the early 1970s and have therefore established their credibility as the dominant protocol for building large-scale and stable networks. Because of the tremendous success of IP, and the lack of any strong need for ATM's quality-of-service guaranties, it is difficult to find fault with packet switching or, more succinctly, to find many tasks on the Internet that ATM does better than IP. Some of the most often cited shortcomings of IP networks are associated with either slow speed or the high prices of routers.

On one hand, ATM advocates claim that the technology does address some of the fundamental limitations of IP routers listed below. These items are directly related to the features that can be provided if the router is combined with an ATM switch.

- Poor scalability to large numbers of physical interfaces per router
- Poor support for very high speed ports such as OC12 and above which are required in enterprise and Internet backbones
- Little support for differing degrees of quality of service and very little experience implementing quality of service in routers

On the other hand, the fundamental limitations of ATM switches, as listed below, are quite different from routers.

- Complex signaling and routing protocol
- Little real-world, large-scale experience
- Poor integration with layer 3 technologies (i.e., cumbersome glue provided in the form of MPOA, LANE, Classical IP over ATM, MARS, etc.)

Even though these protocols have some serious deficiencies, the promising aspect is that, for the most part, the limitations of layer 2 and layer 3 technologies are orthogonal. Therefore, if the two layers could be integrated in a manner where the coupling was clean, efficient, and realizable, the technological product would be a best of breed. The realized combination, network layer switching, is a product that contains the IP routing intelligence of a router with the fast data-forwarding skills of an ATM switch.

By combining the strengths of layer 2 and layer 3 technologies, network layer switching intends to build a system that accomplishes the following:

- Provides fast and low latency over ATM
- Redirects data flows or routes to specially engineered ATM virtual circuits
- Analyzes data with a view towards redirecting these data on traffic engineered virtual circuit.

The problem of designing new protocols that combine layers 2 and 3 in a manner different from the IETF and ATM Forum's current proposals can be divided into two categories: flow-based versus topology-based. From a high-level view of network switching, one can think of the flow-based models as building a network out of router/switching devices in which unique ATM virtual circuits are created for each IP conversation, where a conversation is synonymous with a file transfer or WWW session. In the topology proposals, the router/switching devices use their ATM fabrics to create ATM virtual circuits that can carry all the traffic destined between pairs of subnetworks or IP routes.

A few major vendors have contributed most of the work done in defining these new protocols. At the time of this writing, there are four major protocol architecture proposals, which can be subdivided into two main categories. These two categories are on standards track primarily in the IETF. In order to keep the material in this chapter manageable and of reasonable length, only the leading proposal from each category in the standards process will be described in detail.

In the short term, it is generally believed that only the flow-based technology will be available. The reasons for this, as described in greater detail below, are its head start in the market and its widespread industry acceptance. The next section will address flow-based network layer switching. There is a great deal of controversy over the use of network layer switching paradigms and a great deal of study has gone into the topology-based models. The general consensus is that flow-based protocol will be used in the LAN and topology-based protocols will be used in the WAN. For this reason Section 13.4 describes a leading topology-based proposal, Cisco System's tag switching, in detail. Ultimately, both the flow and topology proposals will be merged into a standard incorporating traits from each.

13.2.2 Network layer switching models

Integration of layers 2 and 3 means that ATM switching is used both as the physical interconnection medium and as a high-speed forwarding technology. IP routing technology is then used to propagate routing tables and determine what path data should follow. If a network designer accepts the above arguments, then a final point to be considered is exactly when ATM switching should be employed. There are two different philosophies regarding how ATM should be used when coupled with IP routers.

The major distinction between the flow-driven versus the topology-driven proposals is the level of granularity applied when considering how IP traffic should be carried over ATM. In some cases, these proposals are known as fine versus coarse-grained switching. The fine-grained nomenclature applies to the flow-based models and the coarse-grained terminology applies to the topology-based proposals. However, regardless of the name used, the principle is the same: they both deal with different abstractions of traffic passing through the network.

When flow-driven IP switching is used to carry user datagrams, the interconnecting IP switches initially employ predefined virtual circuits to interconnect routers carrying the IP default route data flows. In both the flow-driven model and the topology-based model network layer routing intelligence is integrated with the ATM switches' control software. Therefore, each ATM switch has the ability to understand the traffic crossing the ATM switch from a layer 3 perspective and consequently make decisions on how that traffic can be switched as it crosses the ATM network. The IP packet analysis code and the predefined virtual circuits are, in theory, only used for short periods of time. When the routers/switches detect a traffic pattern that triggers the creation of a separate virtual circuit for the datagrams, the IP control software on the routers/switches is no longer needed to route packets and the ATM hardware's VCI table is used to forward the cells.

The concept of the router, being coupled with a switch, and able to examine all the traffic is actually true in both fine- and coarse-grained aggregation; however, in the fine-grained model the router's interaction may be more common. The reason for this interaction is that the switch must be aware of each individual IP session that is crossing the switching fabric. IP sessions can be defined by several different metrics; however, the most common metrics are combinations such as the IP source/destination addresses coupled with a TCP port number. The port number will correspond to electronic mail, file transfer, or WWW page transfer. Therefore, when the IP switch recognizes a series of packets with the same addresses and port number, it will consider them as a singular flow and create a segregated virtual circuit to carry the subsequent data.

An alternate method for coupling IP routers with ATM switching is done by using topology-based switching, which is also known as coarse-grained aggregation. With this method, the IP router/switch still uses IP's routing protocols to determine where data should be forwarded; however, the major difference is that when traffic is segregated onto new ATM virtual circuits it is not done on a per-flow basis. With coarse-grained aggregation, the data pattern recognized is at the level of IP routes. Therefore, each IP switching router at the edge of an ATM network associates ATM virtual circuits with the other routers at the edge of the network.

The arguments for this approach are that the same benefits of fine-grained aggregation can be achieved; however, because fewer ATM resources are needed, the system can scale to support more attached devices. In addition, some network architects feel that this level of aggregation matches existing routing protocols well and is better suited to traffic engineering.

13.2.3 Network layer switching proposals

As mentioned previously, there are currently four major proposals for network layer switching. These can be subdivided into two main categories—fine and course aggregation. These proposals are all under review in the IETF and several have been released as information RFCs. Informational RFCs are designed only to assist with information dissemination and do not represent IETF standards. However, they can be used to start the process of technology innovation. Ultimately, these various proposals will be studied, and the official standard from the IETF will be presented in the Multiprotocol Label Switching (MPLS) RFC.

Because the IETF's work is still in its infancy, the information provided in the chapter is somewhat broad and may change as the work progresses. What is clear from the MPLS baseline document is that the IETF's tendency is more aligned with the topology-based proposals. However, to date, the vast majority of actual implementations of network layer switching is focused on the flow-based models, primarily because they were the first introduced and documented in RFCs. Therefore, the material in the next section will focus on the most prevalent flow-based proposal on the market, Ipsilon's IP switching. The material following IP switching will present a topology-based network layer switching technology that is closely aligned with the IETF's MPLS baseline, Cisco System's tag switching.

13.2.4 IP switching—fine grain aggregation

IP switch implements the IP protocol stack on ATM hardware, which operates as a high-performance link layer accelerator. The IP switch protocol is designed to dynamically shift between default routing using store-and-forward and cut-through switching based on flow patterns seen in the traffic. The flow recognition is actually much the same as an MPOA edge device. Because IP switches are fundamentally routers running traditional IP routing technology, they integrate easily into existing internetworks. Default routing decisions are based on IP protocols, so IP switches behave like other IP nodes and are naturally interoperable with existing applications and network management tools.

The operation of an IP switch follows the steps of initialization, neighbor peering, default forwarding, and, finally, shortcut creation. The operation within the IP switching system is derived from two published protocols: the Generic Switch Management Protocol (GSMP) and the Ipsilon Flow Management Protocol (IFMP). At startup, each IP switch sets up a virtual channel on each of its ATM physical links to be used as the default forwarding channel. IP data traffic can enter these default ports if passed from an upstream host or edge router equipped with an ATM interface supporting IP switching software. As the IP switch receives incoming traffic from the upstream devices, it uses AAL5 to reassemble the data and examines the IP header via IP routing software on a router processor. The ATM switch hardware functions simply as a high-speed extension of the routing software and as a transport medium for interconnecting IP switches.

After the route for the packet has been determined, the IP switch segments the packet using AAL5 and forwards it to the correct downstream IP switch where the process is repeated. The major difference between the IP switch and a router at this point is that the IP switch records information about the packet header and the direction in which it is forwarded. In this way, the IP switch is attempting to perform the same function of an MPOA edge device, finding flows of similar traffic. (See figure 13.4.)

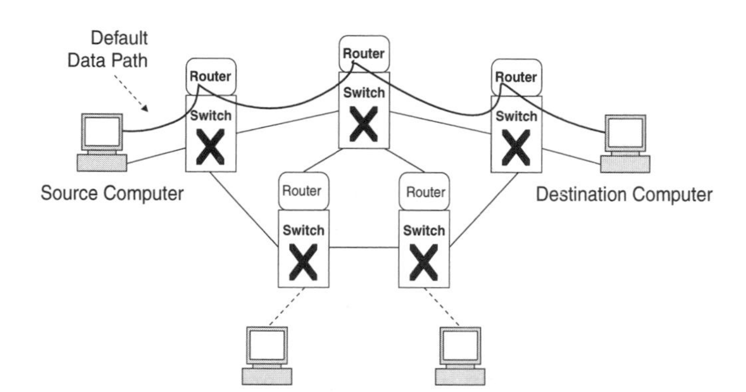

Figure 13.4 IP switch default behavior

From the perspective of the IP switch, a flow is a sequence of IP packets sent from a particular source to a particular destination sharing the same protocol type (such as UDP or TCP), type of service, and other characteristics as determined by information in the packet header. The IP switch controller identifies flows, since cut-through switching in the ATM hardware can optimize these. The rest of the traffic that is not part of the flow continues to receive the default treatment of hop-by-hop, store-and-forward routing.

Once a flow is identified, the switch controller asks the upstream node to segregate the flow using an ATM virtual channel. When the upstream node has created the virtual channel, it diverts the traffic to the new virtual channel. Independently, the downstream node also can ask the IP switch controller to set up an outgoing virtual channel for the flow. When the flow is isolated, the IP switch controller instructs the switch to make the appropriate port mapping in hardware, which bypasses the routing software and its associated processing overhead for all subsequent packets. The cut-through path eventually does timeout and is deleted when data are not being exchanged over the virtual circuit. (See figure 13.5.)

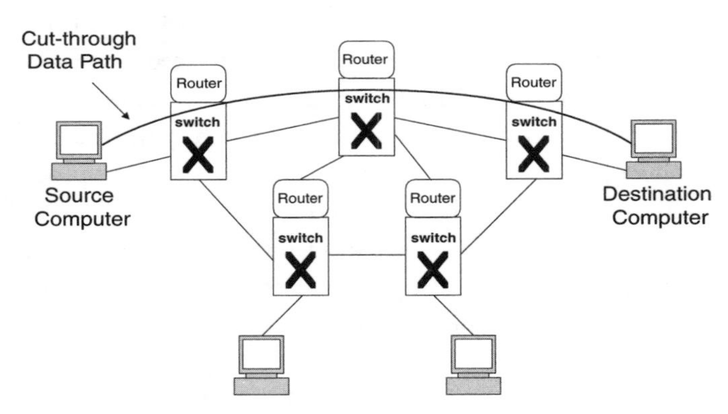

Figure 13.5 IP switches in cut-through mode

This design allows IP switches to forward packets at rates limited only by the aggregate throughput of the underlying ATM fabric. First-generation cell switching routers support up to 5 million packets per second, which is far greater than comparable routers. Furthermore, once the cut-through path has been created and is operational, there is no need to reassemble ATM cells into IP packets at intermediate IP switches. Therefore, throughput remains optimized throughout the IP network.

An advantage of building a system that supports all current major IP protocols is that IP switches integrate smoothly in that environment. QoS requests for each flow are supported using the Resource Reservation Protocol (RSVP), and IP switches can support native IP multicast without any modification to standard multicast protocols, such as the Distance Vector Multicast Routing Protocol (DVMRP) and the Internet Group Management Protocol (IGMP). Because of this smooth integration, IP switches make an excellent building concentration edge device on a campus network in which the core backbone is pure ATM.

Finally, because IP switches are basically optimized flow detection and cut-through devices, their benefit will be best realized with long lived IP streams such as those seen in the HyperText Transmission Protocol (HTTP), Telnet, or multimedia audio/video conferencing. IP switches may not yield any noticeable improvements over traditional routers when the IP stream is composed of protocols such as the Simple Mail Transfer

Protocol (SMTP), Domain Name Service (DNS), or the Simple Network Management Protocol (SNMP).

13.2.5 Multiprotocol and tag switching

The main motivation for tag switching* is to solve networking problems posed by Internet service providers. However, the technology is not restricted to this environment and can be applied in enterprise networks as well. Tag switching addresses a number of the problems facing service providers, such as:

- Increasing the demands for throughput
- Scaling
    - Number of network nodes
    - Number of routes on routers
    - Number of concurrent data flow
- Engineering traffic or traffic control. Making traffic follow known explicit paths to improve performance
- Improving the ability to incrementally enhance the functionality of IP routing on ISP routers
- Simplifying the integration of ATM and IP technologies.

The basic idea of tag switching is to integrate OSI layer 3 (network layers such as IP) control and OSI layer 2 (data link layer such as ATM). The advantage of this is to get the benefits of ATM switching technology, that is, moving data very quickly with optimized hardware, and Internet routing protocols. Therefore, tag switching attempts to glean the best from both technologies. It is important to note that tag switching support can be implemented with only a software upgrade to routers or ATM switches.

As with IP switching, tag switching attempts to use only what are thought to be the best aspects of ATM without implementing the majority of the ATM Forum's specifications. Tag switching specifies how routers can use ATM to communicate between ports, but it does not require the following:

- Mapping of IP address or E.164 address
- Mapping RSVP to ATM switched virtual circuits
- Mapping BGP to PNNI
- Mapping TCP flow control to ABR.

---

* Tag switching is currently progressing towards Internet standard status within the IETF MPLS [5] group.

An additional benefit of tag switching is evident in the ability of protocol designs to make changes in either layer 2 or layer 3 without the change in one impacting the other. There is a subtle distinction that tag switching provides for in integrating these technologies. On the one hand, it tries to improve the interworking between layer 2 and layer 3; on the other hand, it provides enough separation to allow changes to either. Dividing layer 2's forwarding from layer 3's routing/control capabilities accomplishes this.

13.2.6 Tag switching operation

The fundamental building blocks of a tag switching network are conventional routers and ATM switches as shown in figure 13.6. The edge routers are called tag edge router or tag imposition devices. In the center of the network are ATM switches or more traditional IP routers. Within the large complex network there will typically be a subset of devices that implement the tag protocol. However, the devices in the center of the network do not necessarily need to support the tag protocol, because they can forward datagrams using pure ATM. Only the edge devices need to support the protocol, while the edges mediate between the two protocols.

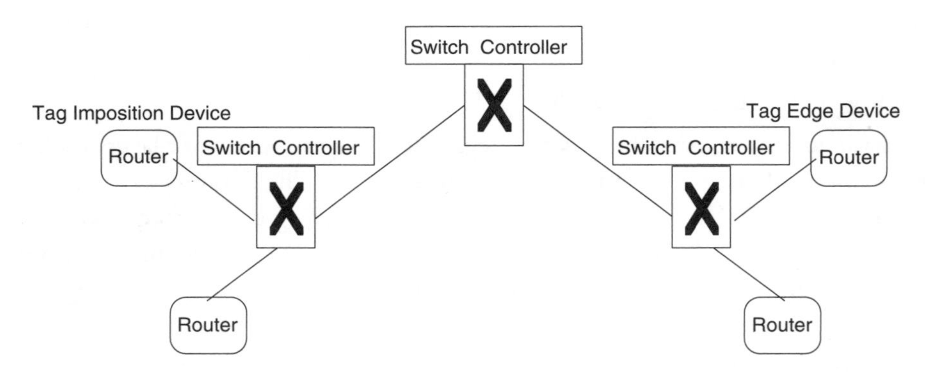

Figure 13.6 Tag switching architecture

Logically, the tag switching protocol is subdivided into two components: forwarding and control. The forwarding component is based on "label swapping," which is basically the same principal employed by ATM switches as they forward cells. That is, the VPI/VCI conversion from the one received on the input port to the VPI/VCI transmitted on the output port. The control component provides the functionality of traditional IP routing. However, it is important to note that the control component can be extended to support protocols beyond IP, because the tagging protocol can make tag associations based on any layer 3 protocol. In addition, to help ISPs, the control component can also be extended to support new features for IP routing protocols.

The tag forwarding component of the system is based on the ability of network devices to acquire the tagging information. In a sense, tag switching can be looked at as a system where edge routers exchange information to the core devices, as shown in figure 13.6, so that the core devices pass packets by examining only the VCI (tag). The core devices, even if capable of understanding IP routing protocols, do not perform routing on a packet-by-packet basis. As with ATM's VPI/VCI, the actual tag numeric values only have significance on a hop-by-hop basis. The value may change as the tag path crosses the network. The path created between tag edge devices typically corresponds to an IP route and therefore provides coarse-grain aggregation.

The core devices receive the command to create inter-edge router path via a Tag Distribution Protocol (TDP). This protocol is used to create the core tag switching device's Tag Forwarding Information Base (TFIB). The process is divided into a phase where the routers boot up and begin to exchange routing information along default paths or virtual circuits. They next recognize routes that exist between themselves and start a phase requesting that special paths be created between the routers. These paths are typically ATM virtual circuits.

The routers forward packets down these new paths by placing the data packet inside a special tag packet. As with ATM, the intervening switches/routers do not need to examine the contents of the cell/packet to forward the data as they cross from router to router. One of the major differences between tag switching and IP switching is that the forwarding decision is based on an aggregation at the level of an IP route, as opposed to per-flow switching.

A Tag Forwarding Information Base is made up of the following fields, which should provide all the information required to forward a packet.

- An incoming tag, which is an index used to quickly find the tag entry in the table
- An outgoing entry from the table containing
    - Output port, the port on the tag switch where the data should next be forwarded
    - Output MAC value, if the output port is a shared media, this value is needed to identify the next-hop router/switch.
    - Output tag value for the next-hop tag switch

When a packet arrives at a tag switching router, the input tag is looked up in the TFIB, the original tag is removed, the new values from the TFIB are placed in the packet header, and then transmitted. In the case of multicast, each incoming entry has multiple outgoing entries.

The Tag Distribution Protocol updates typically run separately from the routing protocol updates and the frequency of advertisements can be low. This is because inter-router TDP sessions run over a reliable transport protocol (TCP) and use incremental updates. When the network is being initialized, there will be a large number of tag update messages; however, once routing is stablilized, there is no reason to send an additional tag update unless the topology changes.

Tag updates over a reliable transport also allow routers participating in TDP to quickly realize when sessions have died and routing updates/changes need to be made. There is a subtle point here that is worth clarifying. When using tag switching without operator intervention, the path the data follows is the same as the IP routing protocol's selected path. Tag switching utilizes routing information at the edges of the network to make its forwarding and path creation decisions. When IP routing changes, then the tag forwarding decisions also change.

Within the routers the TFIB will resemble the routing table with some additional information. This is illustrated in figure 13.7. The table on an edge router will contain IP address prefixes along with output tag values. The output table is a tag advertised by the adjacent router, coupled with the IP prefix. When an IP address packet arrives, the prefix is checked for the longest match in the IP routing table. The process of longest match lookup can be considered as part of tag switching control and only needs to be performed at the boundaries to the tag switching network. This overhead processing of locating not just a path to the destination, but the best path from the routing table, is not needed in the tag switching core.

The best-match value is then used to create the next-hop tag packet. In cases where there are redundant paths, or if the TFIB contains more than one entry, traditional routing protocol decisions will be used to select the correct path.

The tag switching behavior described above is valid when an edge or tag imposition router is transmitting a packet into a tag aware infrastructure. When the packet is inside the network, there are two possible methods of processing the data flow. First, if the interconnecting device is an ATM switch, then the packet is segmented into cells and subsequently passed down a virtual circuit created with the TDP. On the other hand, if the intermediate device is not an ATM switch but is instead another router, which supports tag switching, the intermediate router will recognize that it has received the specially formatted packet. It should then perform a lookup in its TFIB for the output tag and retransmit the packet with the new tag value.

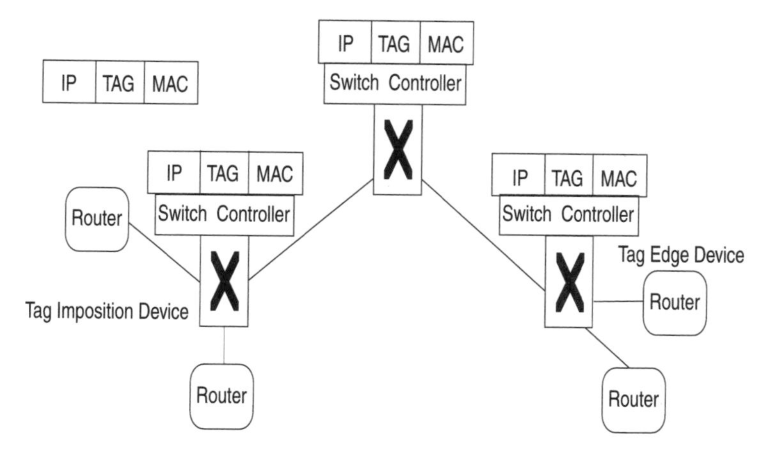

Figure 13.7 Tag forwarding information base example at edge

If the tag switching device is a router, then the tag is found in a special packet type, which has the original packet as the payload. If the tag switching device is an ATM switch, then the tag value is just the ATM VCI.

If the interconnecting device is a router, then the tag table lookup process is very similar to an ATM switch's VC table lookup. Because there is no route computation processing, the packet switching is very fast. It could even be argued that it would be faster to move data with tag switching routers than with pure ATM switches, because data packets are typically larger than ATM cells. Therefore, if the VC or TFIB table lookup process become a bottleneck for high-speed switching, the device that switches larger data units will have better performance, because it needs to access the table lookup fewer times. Conversely, ATM table lookup hardware design can be easily modified to produce higher-performance tag switching engines.

The frame encapsulation for a tag packet is shown in figure 13.8. The encapsulation on frame-based media, regardless of the network layer, is done by prepending a header to the original datagram. The tag header contains the following:

1. The tag value, which is a 20 bit value
2. Class of services is used by service provider to allow for different grades of services.
3. Bottom of stack is used to signify if multiple levels of tag encapsulation are being used, tag switching inside tag switching. This bit would signify when the last level of encapsulation has been reached.

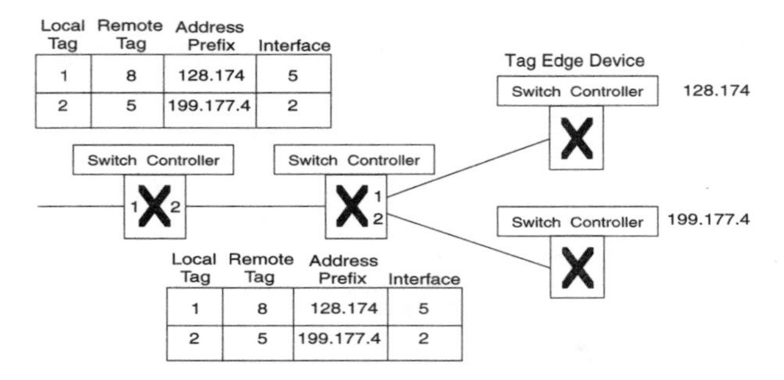

Figure 13.8 Frame encapsulation for a tag packet

4 Time to live: Because packets follow the same path that IP routing has established, has to be a method of detecting whether transient loops exist. If data packets are looping without hope of reaching their destination and should be deleted.

The actual tag packets have their own formats, which Cisco has registered as new ether types* and PPP identifiers. The new ether types allow the interconnecting devices to identify that Ethernet packets do not contain IP headers, for example, but contain tag switching headers running IP. Therefore, there is a unique ether type for IPX, or any other transport layer protocol, running over tag switching. The total data header overhead per tagged packet is one word. One common way to look at this is that tag switching over Ethernet networks is done by placing a "shim" between the network and data link layers.

13.2.7 Issues concerning tag switching over ATM

As described previously, the process of tag switching has been subdivided into two parts, forwarding and control, providing a clear division of labor. Tag switching can be used over virtually any layer 2 technology; however, when run over ATM there are some potential pitfalls. The main difficulty with running tag switching has to do with converting data packets into ATM cells and then correctly mapping the frames onto virtual circuits so that the frames can be correctly reassembled at the destination.

In order to understand the complexities introduced by ATM, refer to figure 13.9. In the figure, two tag imposition devices are transmitting data into an ATM network. If the ATM virtual circuits are naively created to correspond to aggregation of routes, then the two inbound flows are merged onto one output flow using the same VCI field. This produces the undesirable effect of creating packets on the output port that are aggregates

* A designator used to specify the contents of an Ethernet packet.

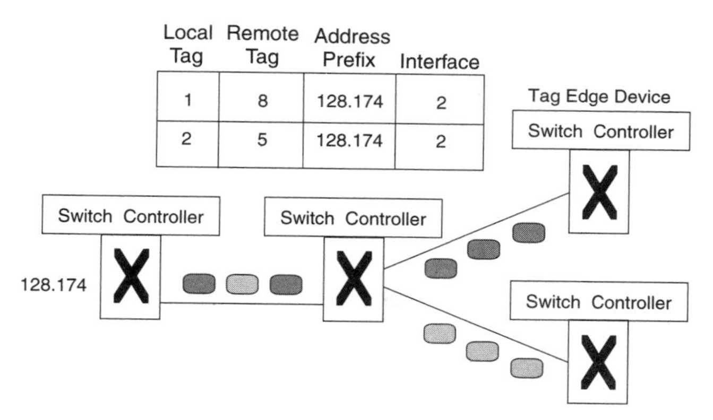

Figure 13.9 Packet interleaving over ATM

of the input packet. When the destination receives one of these double-sized packets, it will not be able to reassemble the contents using the specified AAL.

In order to resolve this problem, two possible solutions have been proposed. In the first method, the ATM switch running TDP realizes that two downstream routers are receiving advertisements for the upstream network. To eliminate the cellmerging problem, the switch allocates an additional tag(s) on the output link, one for each downstream router, that is, the ATM switch assigns a unique tag per interface to each of the downstream tag imposition devices.

By creating end-to-end tag significance, the problem of VCI merge is eliminated in much the same way as end-to-end ATM VCs segregate data flows. However, the tag distribution protocol is still operating on aggregates corresponding to the abstract level of an IP route. An additional drawback is that the number of tag imposition devices per session will grow from N to N^2.

When an interface-specific tag is used on the intermediate ATM switch, the cells from the input ports are still interleaved; however, now the output VCI (tag) values have been changed in their VCI fields. With the unique VCI fields, the cells can then be correctly reassembled at the destination. (See figure 13.10.)

One drawback to this solution is the potential reduction of network scale, because the number of required tags per address prefix needs to be increased; in some topologies the increase can be dramatic.

The second method for coupling with ATM cell interleaving is to force the ATM switch to correct the problem by transmitting only one frame at a time on the output link. This technique is known as VC merge.* The switch can perform this function,

* VC merge is the one exception to the statement that tag switching can be implemented with software only. In the case of VC merging, hardware modifications are required.

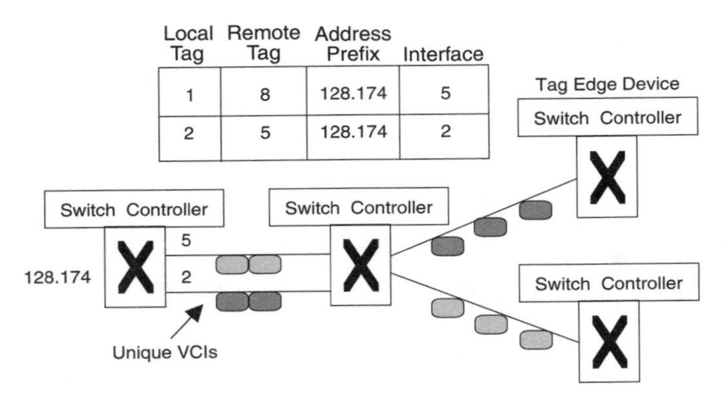

Figure 13.10 Interface-specific tags

because it can look for AAL5 end-of-message flags to determine packet boundaries. When it receives a complete AAL5 frame, it ensures, via buffer management, that it does not interleave the cells. The processing involved with the algorithm is similar to the analysis done by current ATM hardware performing early packet discard, so implementing the feature should not be difficult.

There are drawbacks to this technique since it changes the end-to-end latency performance realized on the VC merging ATM network to act more as a traditional packet-switched network. The latency added is not huge and may be as low as the time required to buffer one packet at OC3c speed. Even if the ATM switch supports a number of port speeds, it will wait until the packet has been completely received into the input buffer before committing to transmitting it on the VC merged output port. In addition, the data are arriving from a packet network, so introducing a protocol that yields packet network performance seems irrelevant.

The overhead of VC merge will require hardware modifications. However, this approach will yield a much more scalable network, because the system only needs to support one tag per prefix or one tag per exit router. In addition, the ATM switch can run this protocol independent of the VC merge algorithm. Therefore, the CBR/VBR, and even ATM signaling, can continue to run on the ATM switch supporting tag switching.

13.3 An alternative to MPOA: ATM API

The second area where techniques other than MPOA may play a key role is the deployment of application program interfaces (APIs), which allow applications to directly communicate with an ATM network to utilize QoS. In order for applications to employ QoS, they may either use a system such as RSVP on an MPOA system, or communicate their traffic characteristics directly to the network. The choice of which technique to use is left purely to the application developer. In this scheme, bypassing RSVP would primarily be done to improve efficiency for special purposes.

APIs in widespread use today, such as UNIX sockets and Microsoft Winsock [3], were not designed with ATM's signaling and QoS in mind. Developers accustomed to working with those APIs would rather not contend with ATM-specific issues when coding new applications. Allowing this requires that a new application program interface will be needed, along with new applications designed to use the QoS API.

Currently, the work of defining a common QoS-capable API to support real-time traffic and resource reservations is beibg done primarily in the Winsock Forum, a coalition of several hundred vendors and developers, and the ATM Forum. This work is planned as a basis for application developers to write programs that use QoS. The ATM Forum's SAA working group is standardizing a set of native ATM service semantics, which will serve as guidelines. In addition, several operating system vendors have begun to implement ATM APIs.

13.3.1 ATM Forum's involvement with APIs

The SAA working group's standard for native ATM services provides standard semantics, which can be applied to the design of APIs. The Forum's proposal provides detailed semantics for setup, use, and teardown of virtual connections between applications. The connection setup semantics include provisions for negotiating connection attributes, such as classes of service, bit rates, and quality-of-service parameters.

The ATM Forum's work specifies three types of Service Access Points (SAPs), which are used to define the interface between layers of the protocol stack.

1. Type 1 SAPs contain ATM addresses concatenated with a data link layer Ethernet or Token Ring address. These are used to address entities that reside at layer 2 of a protocol stack. By using this kind of addressing, the application using the API will place itself logically at the data link layer as a peer to LANE.
2. Type 2 SAPs concatenate a network layer address with an ATM address. ATM connections set up using this type of addressing carry data from only one network or transport protocol. Classical IP is an example of an entity that uses type 2 SAPs.
3. Type 3 SAPs supply ATM addressing for a single application. This is the type of addressing that will be used for most native ATM applications. With type 3 addressing, an application may request a virtual connection for any other application or applications for which it knows the ATM address.

13.3.2 Winsock forum standards

The intent for the resulting solution will be a multiprotocol LAN internetwork scheme, which scales in size and better leverages the capabilities of the ATM fabric. Applications making use of the new API, coupled with a network that supports QoS, will have successfully migrated away from the traditional shared media paradigm. To date, the boldest step towards a true ATM-aware API is the joint effort by Microsoft and Intel Corporation on Winsock-2, follow-on to the de facto standard Winsock API. This API has had its first release for both Windows NT and Windows 95.

Applications written to Winsock-2 will be able to specify QoS parameters, which means that applications can request the bandwidth and latency they need. In addition, they can request point-to-point and point-to-multipoint connections, and use ATM addresses. The first release will rely on AAL5 for segmentation and reassembly into ATM cells, but future releases of Winsock-2 may use other AALs. The API is designed to support a variety of protocols, including ATM, DECnet, IPX/SPX, ISDN, and TCP/IP.

Winsock-2 enables applications to specify QoS parameters for a connection according to procedures defined in IETF RFC 1363, *A Proposed Flow Specification*. RFC 1363 is actually one of the baseline documents that RSVP developers have used. The RFC specifies that when a connection is established between two applications, the source must supply two sets of flow specifications, one for each direction of data flow. The parameters the application must include in the flow specifications are traffic description, latency, and level-of-service guarantee, which all are directly compatible with ATM's QoS parameters specified in UNI 3.0 and 3.1. It accomplishes this using a *root-and-leaves* model of multipoint communication.

Winsock-2 consists of three specifications, an API, service provider interface, and protocol-specific annex, are closely related to the operation of a Microsoft OS. The API describes the commands a software developer can use while programming to the network interface; these are generic to any operating system. When an application is written using the Winsock-2 API, it speaks directly to Winsock-2's Dynamic Link Library (DLL), which provides software functionality on an as-needed basis. The service provider below the DLL describes interworking within the operating system that provides the communication path to the transport protocol, such as TCP. Finally, the third specification, the protocol-specific annex, furnishes access to protocol-specific functionality that documents new features like QoS. (See figure 13.11.)

Similar to the integrated services work in the IETF, the Winsock-2 API defines three levels of service: guaranteed, predictive (controlled load), and best effort. Guaranteed service is designed for applications that require known quality-of-service parameters. Predictive service is designed for applications that have specific needs but

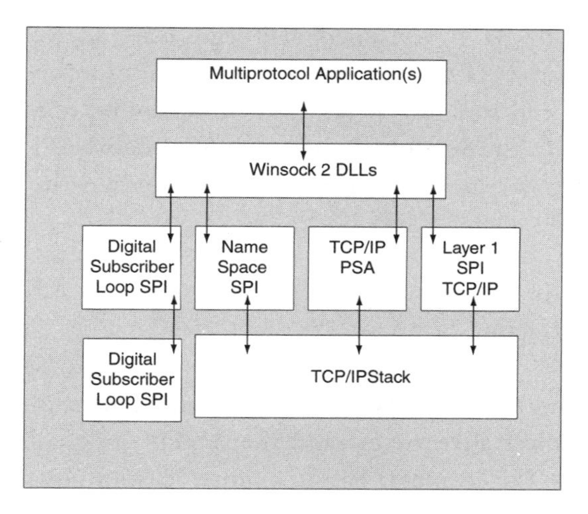

Figure 13.11 Winsock interface

can accommodate or adapt to some variation in service quality, most likely via a feedback mechanism. Finally, best effort provides no guarantees and is intended to provide QoS like today's Internet.

The Winsock-2 API issues commands to the operating system, which then interfaces to the ATM services. When a new switch virtual circuit is needed, the path is created using the parameters specified in the Winsock system call. The parameters are used to construct RSVP's PATH and RESV messages. If the network can't offer the level of service an application needs, the call setup will fail and the application can be notified of the problems. The application can then request a lower quality of service or ask for none at all. When a connection has been successfully established, the application is notified and it also can receive subsequent QoS change notifications as new PATH/RESV messages arrive.

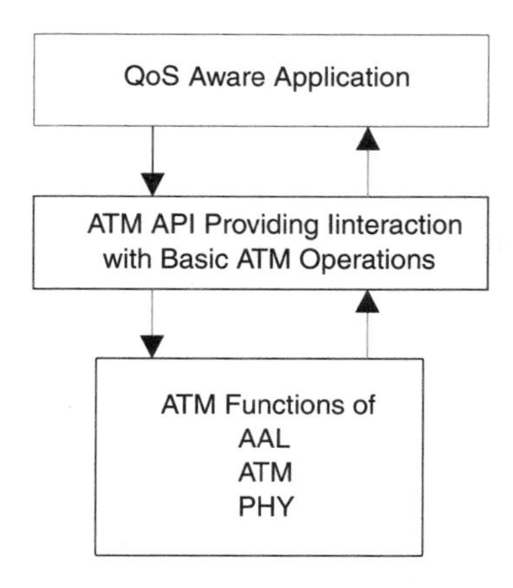

Figure 13.12 Native ATM application

This situation allows hosts to communicate in a QoS-rich network; however, the problem of destination discovery (ARP) remains. This could be solved with LANE/NHRP in which the Winsock application is used in conjunction with the ATM Forum protocols, or Winsock could be used with an ATM DNS-like service. With the DNS service, the host would request from a name server the ATM E.164 address of the end station using only the host name (i.e., www.atmforum.com). In this case the protocol stack would be reduced to that shown in figure 13.12.

It is not clear at this point how quickly ATM developers will begin using native ATM APIs. The important point is that native ATM APIs would allow applications to request the network to signal a particular end station and set up a virtual connection between one application and another. Finally, the difficulty in using the Winsock-2 API will be very similar to Winsock; therefore, developers should be able to rapidly write new applications.

13.4 Outstanding issues

Before concluding this chapter, a brief review of some outstanding issues would be helpful in light of the migration paths "technical alternatives" and the MPOA protocol model. The intent here is not to claim that the "emperor has no clothes," but to illustrate some of the potential problems the industry must face. At this point, these issues are areas of potential concern, and network managers will need to gain experience before coming to final conclusions about ATM applicability. There is no way to predict the outcome of these competing technologies, and most likely, all will be implemented to some degree in enterprise networks.

Some of the more serious concerns regarding ATM are related to its connection-oriented nature and system scalability. These issues are often topics addressed during ATM Forum meetings when members compare ATM to traditional IP routed networks, which have the nice property of aggregation of both routes and traffic. Unfortunately, the internal debates are not often discussed outside the Forum. Some interesting points for network managers to be aware of include the following problems:

- Virtual circuit starvation: This problem is directly related to the fact that ATM is connection-oriented and, when widely deployed, each TCP/IP conversation may require a unique virtual circuit. The potential problem occurs when a network grows very large and the interconnecting ATM Network Interface cards (NICs) and ATM switch ports are taxed. In this case there is a definite potential that the ATM switch and NICs will run out of virtual circuit entries supported in their caches. Even if they could maintain the logical cache entries, they would not be able to physically process all the data arriving on hundreds of concurrent circuits. This problem may only get worse as the ATM network grows nationally or internationally. Because of the nonlocality* of Internet traffic, the virtual circuit space could be exhausted.

---

* The Internet is considered to be a network in which traffic has some locality, but to a certain degree an individual user may transmit data anywhere in the world at any time. Typical users will send most of their packets to local resources (file servers), but when running Internet Browser the locality quickly vanishes.

- State explosion: This problem is also associated with connection-oriented networks. The state explosion problem is due to the requirement of connection-oriented networks to maintain some knowledge of where to route data on a per-virtual-circuit basis. IP routers perform this process on an aggregate basis. State explosion is the potential product of any network that requires some state to be maintained per flow, and it is a concern with RSVP run over a pure IP routed network. A related concern with protocols such as RSVP and MPOA, which establish QoS by maintaining state, is that once these protocols are in place, there is little the network manager can do to keep every end user from running them. In the long run, the network may reach a state where all applications run RSVP, the resources are once again starved, and even RSVP-capable applications experience poor QoS. The analogy to RSVP is often a telephone network. A telephone network works well for mostly local calls and a few long-distance calls; however, the Internet completely changes that model in that long-distance resources may be the most commonly accessed. RSVP policy is an area for future study.
- Multicast explosion: This problem is related again to the number of virtual circuits required to support an application, in particular, multiparticipant multicast. When these applications expand on pure ATM networks, there is a potential that the number of required virtual circuits could become huge, because each sender usually establishes a virtual circuit to each receiver. This problem is solved by placing aggregation devices in the ATM network that combine multiple virtual circuits into one common stream. However, the aggregation process is, in essence, converting ATM into the same network that would have been achieved if switches were never used and all multicast traversed IP routers.

There are some high points to ATM worth considering in light of the preceding discussion. ATM has some clear strengths over traditional IP-based networks.

- The most obvious point is the high speeds that ATM switch ports support. These speeds have traditionally been several times faster than router ports and the difference may not disappear in the near future.
- ATM switches also scale far better than routers in the number of ports per switching element. By current standards, the largest routers have 16 OC12c ports, while ATM switches may support more than 300.
- ATM provides superior tracking and billing features compared to connectionless networks. While this feature is not critical today, it may become increasingly important as the Internet is used for traditional pay-per-use applications, such as voice.

- ATM's connection-oriented behavior provides excellent control over a critical network resource, bandwidth. With the segregation provided by virtual circuits supporting different QoS values, all dividing one physical link, the network manager has tremendous control over how much bandwidth routers can use, down to a per-application basis.
- Finally, packet-switched technologies have traditionally had difficulty supporting isochronous bit rate. This service, which can be used to support circuit emulation has yet to be realized efficiently on IP routers.

13.5 Summary of migration strategies and alternatives

This chapter has covered some of the issues involved when migrating a legacy network to one that supports MPOA. We also discussed some of the alternatives to MPOA, cell switching routers and a native ATM API, in addition to some areas of concern.

While the future of computer communication can never completely be predicted, there are some clear possibilities. First, there is a signal from the industry that the complexity of the ATM Forum documents is becoming an impediment for implementation. This is evident from the development of IP switching, in which only a small, best-of-breed portion of ATM's specifications is used. Because of this new IP switching development, a battle will be fought between cell switching router technology and the ATM Forum's technology, MPOA/NHRP, which attempts to solve the same problems.

One last related point to this challenge would be the standardization process of multi-protocol label switching technology in the IETF. If the IETF can quickly generate a high-quality specification, there will be some very strong alternatives to using MPOA. The likely outcome is not clear; however, customers are interested in solving problems with simple, cost-effective technology. The winner will be the technology with the fastest speed that delivers the most bits per dollar.

Another battle that will take time to resolve is the dispute over connection-oriented versus connectionless networks. To a certain degree, these arguments are philosophical, but some credit must be given to those individuals who argue that ATM's connection-oriented architectures will not scale globally and the only known technology that will is IP on connectionless routing.

The most likely outcome in the near future will be continued pockets of ATM technology that never leave an administrative domain, that is, a core ATM network with the handoff between domains being pure IP/BGP-based. In addition, because RSVP has the blessing of the IETF, it may become the major standard for QoS requests for both the

Internet and ATM networks. The real unknown is what makes more sense: overengineering the network to stay ahead of the bandwidth demand, but keeping the end nodes simple, versus RSVP with connection-oriented resource reservation complexities in hosts.

Finally, there will be a debate against using MPOA/ATM technology in campus networks and instead selecting gigabyte Ethernet switching. This argument carries some credence, because most network managers are very comfortable with Ethernet and, for the most part, the technology is plug and play. ATM, on the other hand, is new and radically different from broadcast media and therefore has a major psychological hurdle to overcome before network managers will use it.

For this reason, it seems that Ethernet may continue to be the dominant technology in the campus enterprise networks and will connect the major share of desktop/building networks. ATM's role will be restricted to the specialty job of a relatively small high-speed backbone hub interconnecting core routers. ATM will, however, continue to play a key role in carrier/ISPs backbones and is the preferred technology for Internet Exchange Points.

The ATM Forum members are extremely aware of these concerns and take a very aggressive role in developing implementation specifications to address them. The first version of the MPOA specification is an excellent example of that effort and has solved many of the problems in its charter. Compliant products have begun to appear on the market and prototype networks have been built. The next step for the protocol will include the enhancement for redundancy and possibly modifications to fix problems realized in the prototypes. Finally, as network managers gain experience building networks with cut-through routing and QoS, software developers can feel comfortable in producing new applications that make use of the MPOA environment.

13.6 References

1. M. Laubach, *Classical IP and ARP over ATM*, RFC 1577, January 1994.
2. Marshall Rose, *The Open Book*, Englewood Cliffs, NJ: Prentice Hall, 1992.
3. G. Armitage, *Support for Multicast over UNI 3.0/3.1-based ATM Networks*, 02/23/1996, *ftp://ietf.cnri.reston.va.us/internet-drafts/draft-ietf-ipatm-ipmc-12.txt.*
4. *http://www.intel.com/IAL/winsock2.*
5. R. Callon, A. Viswanathan, *A Proposed Architecture for MPLS*, draft-ietf-mpls-arch-00.txt, August 1997.

index

D

E

F

G

N

O

P

Q

R

S

T

U

V

W

X